Obese Humans and Rats

Originally published in 1974, this volume examines the behavioural similarities of obese humans and animals whose so-called feeding centre (the ventromedial hypothalamic nuclei) has been lesioned. Both the obese human and the VMH-lesioned animal seem to share a hyposensitivity to the internal (physiological) cues to eating and hypersensitivity to external cues associated with food. Beginning with a review, these obese animals and the human obese are compared point by point on experimental results reported in the literature. Then, new findings are presented that specifically tested humans for relationships that are well-established for lesioned animals. Next, a theoretical framework integrates the human and animal data to postulate that the relationship of cue prominence and probability of response is stronger for the obese than for normal. The causes for this, and the extension of the basis for the obese's eating behaviour to other areas, are discussed in light of further experiments that will make this invaluable reading for all concerned with the history of obesity and the issues of regulatory behaviour.

Obese Humans and Rats

Stanley Schachter and Judith Rodin

Psychology Press
Taylor & Francis Group
LONDON AND NEW YORK

First published in 1974
by Lawrence Erlbaum Associates

This edition first published in 2015 by Psychology Press
27 Church Road, Hove, BN3 2FA

and by Psychology Press
711 Third Avenue, New York, NY 10017

Psychology Press is an imprint of the Taylor & Francis Group, an informa business

A Library of Congress record exists under ISBN: 0470756799

ISBN: 978-1-138-82916-9 (hbk)
ISBN: 978-1-315-73768-3 (ebk)

OBESE HUMANS AND RATS

BY STANLEY SCHACHTER and JUDITH RODIN

With the collaboration of
Patricia Pliner, Lee Ross
William Johnson, Lucy Friedman
Donald Elman, and C. Peter Herman

LAWRENCE ERLBAUM ASSOCIATES, PUBLISHERS
1974 Potomac, Maryland

Lawrence Erlbaum Associates, Publishers
12736 Lincolnshire Drive, Potomac, Maryland 20854

Distributed solely by Halsted Press Division
John Wiley & Sons, Inc., New York

Library of Congress Cataloging in Publications Data

Schachter, Stanley, 1922-
Obese humans and rats.

(Complex human behavior)

Bbiliography: p.
1. Corpulence — Psychological aspects.
1. Rodin, Judith, joint author. II. Title.
[DNLM: 1. Behavior. 2. Behavior, Animal.
3. Obesity. 4. Rats. WD210S291o 1974]
RC628.S36 616.3'98'0019 74-7179
ISBN 0-470-75679-9

Printed in the United States of America

CONTENTS

ACKNOWLEDGEMENTS

Terry Powley and Leon Festinger read early drafts of this manuscript and we owe a considerable debt to them both. Powley was extraordinarily generous in sharing his encyclopedic knowledge of the hypothalamus and eating behavior. Festinger knows nothing of the hypothalamus but his critical and editorial skills played a major role in shaping this book.

Nancy Goodman and Deborah Perlick were most helpful in checking tabular material and references in the Appendix. Peggy Humphreys, Elaine Gaylard and Claire Shindler typed the manuscript.

Most of the research was supported by National Science Foundation grants GB 29292 to Schachter and GS 37953 to Rodin. The American Psychological Association granted permission to reproduce portions of articles that originally appeared in its journals.

1
BEHAVIORAL SIMILARITIES OF THE VMH-LESIONED ANIMAL AND THE OBESE HUMAN

Of all human frailties, obesity is, perhaps, the most perverse. The penalties are so severe, the gratifications so limited, and the remedy so simple that obesity should be the most trivial of aberrations to correct. Yet, it is among the most recalcitrant. Almost any fat person can lose weight; few can keep it off.

In the search for the causes and correlates of this perverse condition, perhaps the most dramatic fact yet uncovered is that it is possible to produce immensely fat animals by brain lesioning techniques. Bilateral lesions in the ventromedial nuclei of the hypothalamus (VMH) will produce an animal that, for a time, will eat huge quantities of food and grow extraordinarily fat. Classic descriptions (Brobeck, 1946; Hetherington & Ranson, 1940) of the lesioned animal present a creature that staggers over to the food hopper immediately after its operation and begins shoveling in food. For several weeks this voracious eating continues, and there is, of course, very rapid weight gain. This is called the dynamic phase of hyperphagia. Finally, a plateau is reached, at which point the animal's weight levels off, and its food intake drops to a level only slightly above that of the normal animal. This is called the static phase. During both the static and dynamic stages, the lesioned animal is also characterized as markedly inactive and is finicky, irascible, emotional, and generally bitchy. This general hyperphagic syndrome has been demonstrated on rats, cats, mice, geese, monkeys, rabbits, goats, dogs, pigs, and sparrows.

In contrast to the intact animal, the lesioned animal appears to be insensitive to the physiological cues associated with food deprivation. It is, however, remarkably sensitive to the cue properties of its diet, for it will eat large quantities of tasty food but almost nothing of unpleasant food—an eating pattern also characteristic of the obese human. A variety of studies have indicated that the eating behavior of the human obese subject is also relatively independent of bodily needs but is in good part determined by the cues that are associated with food and with the eating routine and ritual. In contrast to normal subjects, the amounts eaten by the obese are markedly affected by such variables as the sight of food, its quality, the time of day, and so on.

The similarity of these findings to those obtained with the laboratory animal made obese by lesioning techniques has been noted by Mrosovsky (1971), Nisbett (1968a, in press), Penick and Stunkard (1970), Schachter (1971a, 1971b) and a variety of other investigators. Both the VMH-lesioned rat and the obese human appear to share a pattern of hyposensitivity to the internal or physiological cues associated with eating and hypersensitivity to the external cues associated with food proper. Though these similarities are, on the surface, striking, the attempts to push the parallel may simply be capitalizing on an engaging, occasional resemblance between two otherwise remotely connected sets of data. It is the purpose of this monograph to explore this parallel in detail. We shall first review the existing experimental literature on lesioned animals and obese humans, in order, where sensible, to make a point-for-point comparison of the two. Second, we shall present the results of a series of experiments specifically designed to test, for humans, relationships that have been well-established for lesioned animals. Next, we shall present a schema and related experimental work which attempt to integrate the various data, animal and human, into a single conceptual framework. Finally, we shall present a series of experiments designed to examine the extent to which our findings on the eating behavior of the obese generalize to nonconsummatory behaviors.

EXISTING PARALLELS BETWEEN THE VMH-LESIONED ANIMAL AND THE OBESE HUMAN

Since making comparisons between animals and humans can be a particularly tricky business, we should like at this point to be completely explicit about the ground rules that have guided this effort.

First, to make sure that we are not simply pointing up an occasional, catchy similarity, we have attempted to learn *every* behavioral fact about the lesioned animal and about the obese human. To do so we have, for animals, relied on the major texts and treatises on eating and the hypothalamus to guide us to the older work. We have used *Psychological Abstracts* and *Index Medicus* from 1960 through 1967 to guide us to the more recent research literature. From 1967 through early 1971 we have covered all issues of the *Journal of Comparative and Physiological Psychology* and of *Physiology and Behavior*. Despite these efforts, we have no question that there is research that we have overlooked. We are, however, confident that our efforts have been systematic enough so that we will not misrepresent what is known.

For humans we have not made a systematic library search. We have both been involved in this area for so long that we believe we know the relevant experimental literature.

Second, we have been repeatedly faced with the problem of deciding what is an unequivocal, well-established fact[1]. That scientific bugaboo—failure to replicate—has complicated this as it does all similar "state of the art" surveys. Rather than attempt to reconcile or explain away incompatible findings we have adopted a batting-average approach. For every fact with which we will be concerned we shall simply report the proportion of all relevant studies that have worked in a particular direction. This procedure opens up the final question—which studies should be included in the batting averages? We adopted the following criteria:

(*a*) Where possible, we include only studies based on behavioral data. For animal work, of course, this decision causes no problems. For humans, we include the results of surveys or clinical papers only when there are no relevant experimental data. With all respect to these methodologies, we are flatly skeptical of the self-reports of fat people about how much they eat or exercise.[2]

(*b*) We include only studies that report all or a substantial portion of the data, rather than report the author's impressions or present only the data of one or two presumably representative cases. Along this same line, we should note that in all cases we have relied on the data, not on what an experimenter said about his data. It may seem silly to make this point explicit, but in a few studies, for some perverse reason, the experimenter's conclusions simply have nothing to do with his data.

We note explicitly that the use of inferential statistics is not one of our criteria either for inclusion in our batting averages or for judging the reliability of a reported relationship. Many of the animal studies, particularly the early ones, do not use statistics. Many of those that do, so clearly misuse a technique as to make their analyses invalid. The most common error is the use of the trial rather than the animal as the unit of analysis (*N* is calculated as number of animals times average number of trials)—a violation of the assumption of independence which illegitimately inflates *N*. Short of going back to the original papers, we would suggest that the reader judge the "solidity" of a reported relationship by its batting average, i.e., the consistency with which it is reported in the literature.

Finally, it should be noted that throughout this exercise, we will consider the data only for animals in the static phase of obesity, animals who, like our

[1]Even the apparent fact that a ventromedial lesion produces a hyperphagic, obese animal has been questioned—see Reynolds (1963), Rabin and Smith (1968), and Han, Lin, Chu, Mu, and Liu (1965).

[2]In three of four questionnaire studies of eating habits (Beaudoin & Mayer, 1953: Johnson, Burke, & Mayer, 1956; Ross, Pliner, Nesbitt, & Schachter, 1971; and Stefanik, Heald, & Mayer, 1959) fat people report eating considerably less food than do normals. In contrast, we note Stunkard's report (personal communication) of his series of chronic fat patients who were fed *everything* that, in interviews, they admitted to eating daily, and who all steadily lost weight on this diet.

human subjects, are already fat. In general, however, the results for dynamic and static animals are quite similar. In Chapter 11, we shall specifically compare dynamic and static animals, as well as currently and formerly obese humans.

Let us turn now to those relationships which have been studied in both the hyperphagic animal and the obese human.

Finickiness

Though the lesioned animal is a heavy eater, it is also a picky one—peculiarly sensitive to the effects of the texture or taste of its diet. If quinine is added to its food, it drastically decreases its intake to levels far below that of a normal animal's whose food has been similarly tainted. On the other hand, if to its normal food is added dextrose, lard, or something that is apparently tasty to a rat, the lesioned animal increases its intake to levels considerably above its regular intake and above the intake of a control rat whose food has also been enriched. A summary of the effects of good food on intake is presented, for animals, on the left-hand side of Table 1. The column labeled "Batting average" indicates that we found six relevant studies and that all six found that lesioned, static, obese animals ate more of a good-tasting food than did their normal controls.[3] In this and all other tables in this chapter, the specific studies involved are identified at the bottom of the column. The Appendix to this volume describes the relevant details of each of these studies, identifies precisely the data on which these analyses are based, and describes the results of several studies which, though possibly relevant, are, for one reason or another, not included in these analyses.

The column labeled "Mean O/N" is our shorthand method of indicating the direction of an effect and providing a crude notion of the magnitude of the effect. For each study we calculate an obese to normal (O/N) ratio by simply dividing the magnitude of the effect for obese subjects by the magnitude of the effect for normal subjects. Thus, if in a particular study, fat rats ate an average of 15 grams of food and normal rats ate 10 grams, the O/N ratio would be 1.50. In Table 1, it can be seen that the average O/N ratio for the six studies concerned with the effects of taste on consumption is 1.53, in-

[3]The specific figures in this and several of the other tables in this chapter differ slightly from corresponding figures in preliminary published reports of this work (Schachter, 1971b). These slight discrepancies are due either to the inclusion in the present tables of work which has appeared since the publication of the first report or to our discovery of relevant papers of which we had not known at that time. In addition, we were able to learn new information about a few studies which either made them irrelevant to a particular tabulation or permitted us to calculate more accurately some of these ratios. Finally, in the summary of studies on the effects of taste on food consumption, we have not included those concerned with the effects of adulterated water—some of which were included in the preliminary report of this work.

TABLE 1
The Effects of Good-Tasting Food on Consumption

Animals		Humans	
Batting average	Mean O/N	Batting average	Mean O/N
6/6	1.53[a]	3/3	1.37
Studies[b]		Studies	
Carlisle & Stellar	(1969)	Decke	(1971)
Corbit & Stellar	(1964)	Nisbett	(1968b)
Graff & Stellar	(1962)	Nisbett & Gurwitz	(1970)
Lipton	(1969)		
Miller, Bailey, & Stevenson	(1950)		
Teitelbaum	(1955)		

[a]This mean does not include the data from Miller et al. (1950), since they do not report raw data for their unadulterated, high-fat diet but indicate only that their lesioned animals outate the normals on this diet.

[b]We include only studies which have added a good tasting substance to the animal's regular food. Thus, the results of Maller's (1964) experiment which added saccharine or glucose to water are not included. Contrary to the consistent results with food, lesioned animals drink far less sweetened water than do normal control animals.

dicating that fat rats on the average eat 53% more of good-tasting food than do normal rats.[4]

Data, from what we believe to be comparable studies conducted on humans, are presented on the right side of Table 1. These studies were designed to compare the effects of good- and bad-tasting food on the consumption of obese and normal subjects. The good foods were such things as vanilla ice cream or milk shakes. In all studies, fat subjects ate more good food than did normals; the average O/N ratio is 1.37, indicating that obese humans eat 37% more of good food than do normally-sized humans.

[4]Admittedly, averaging ratios in this fashion is at best a questionable procedure, for it gives equal weight to studies with differing numbers of cases, takes no account of procedural differences among the studies averaged, and so on. We use this average as a convenient device for conveying some rough idea of the magnitude of a phenomenon. Having monkeyed around with a variety of other devices for conveying this information, we are convinced that in no case does this mean ratio seriously distort the drift of the data it summarizes.

TABLE 2
The Effects of Bad Tasting Food on Consumption

Animals		Humans	
Batting average	Mean O/N	Batting average	Mean O/N
3/4	.76	1/2	.84
Studies[a]		Studies	
Graff & Stellar	(1962)	Decke	(1971)
Hamilton & Brobeck	(1964)	Nisbett	(1968b)
Miller et al.	(1950)		
Teitelbaum	(1955)		

[a] Again, we consider only studies in which the unpleasant substance has been added to food, not water. Corbit (1965) and Nachman (1967) have added quinine to water, and consistent with the general pattern of results with food, they find that lesioned animals drink less. The effect is quite strong; the mean O/N ratio for these two studies is .20.

Table 2 summarizes the results of studies concerned with the effects of bad taste on consumption. For both animals and men, in all of these studies bad taste was manipulated by the addition of quinine to the food. There are four animal studies; three of the four indicate that fat animals eat less than normals, and the average O/N ratio is .76. There are two human studies: One of the two indicates that the obese eat considerably less bad food than do normals; the other finds no significant difference between the two groups, and the mean O/N ratio for these two studies is .84. For this particular fact, the data are more fragile than one would like, but the trend for the two species is similar.

Eating Habits

The eating habits of lesioned rats have been thoroughly studied, particularly by Teitelbaum and Campbell (1958). It turns out that static, obese rats eat on the average slightly, not considerably, more than normal rats. They also eat fewer meals per day, eat more per meal, and eat more rapidly than do normal animals. For each of these facts, there exist parallel data for humans.

Considering first the average amount eaten per day when on ad-lib feeding of ordinary lab chow or pellets, the figures presented in Table 3 indicate that

TABLE 3
Amount of Food Eaten Ad Lib

Animals			Humans	
Batting average	Mean O/N		Batting average	Mean O/N
9/9	1.26		2/3	1.16

Animals: Studies		Humans: Studies	
Brooks, Lockwood, & Wiggins	(1946)	Ross	(Chapter 7, this volume)
Corbit	(1965)	Schachter & Friedman	(Chapter 2, this volume)
Ferguson & Keesey	(1971)	Schachter & Gross	(1968)
Graff & Stellar	(1962)		
Hamilton & Brobeck	(1964)		
Hetherington & Ranson	(1942)		
Nachman	(1967)		
Teitelbaum	(1955)		
Teitelbaum & Campbell	(1958)		

consistently, static obese rats eat somewhat (26%) more than do their normal counterparts.[5] The data for humans are derived from all of the studies we know of in which eating is placed in a noshing or ad-lib context; that is, a bowl of ordinary food, usually nuts or crackers, is placed in the room, the experiment presumably has nothing to do with eating, and the subject is free to eat or not, as he chooses, just as a rat is in its cage. In two of the three experiments conducted in this context, obese subjects eat slightly more than normals do; in the third experiment, the two groups eat precisely the same number of crackers. For both humans and rats, then, the fat subject eats only slightly more than the normal subject.

Turning next to the number of meals per day, Table 4 indicates that for both rats and humans, the fatter subjects consistently eat somewhat fewer meals per day. A rat meal is defined by Teitelbaum and Campbell (1958) as

[5]Virtually every study on the VMH-lesioned animal presents data on the amounts eaten. Since an enormous number of studies are involved, we have not attempted to make an exhaustive listing in Table 3 but simply present a representative sample of such studies. We know of no study which does not support the conclusion that the static obese animal eats somewhat more of freely available lab chow or pellets than does its normal counterpart.

TABLE 4

Number of Meals per Day

Animals: Batting average	Animals: Mean O/N	Humans: Batting average	Humans: Mean O/N
3/3	.87[a]	3/3	.92[b]

Animals: Studies		Humans: Studies	
Brooks et al.	(1946)	Beaudoin & Mayer	(1953)
Larsson & Strom	(1957)	Johnson et al.	(1956)
Teitelbaum & Campbell	(1958)	Ross, et al.	(1971)

[a] This mean does not include data from the Larsson & Strom study, since they report only the range of number of meals eaten for each group.

[b] This mean is based on data only from the Beaudoin and Mayer and Ross et al. studies. The Johnson et al. study presents its data in such a way that though one can deduce that the obese must have eaten fewer meals and snacks per day, it is impossible to compute an O/N ratio.

TABLE 5

Amount Eaten per Meal

Animals: Batting average	Animals: Mean O/N	Humans: Batting average	Humans: Mean O/N
2/2	1.34	6/6	1.24

Animals: Studies		Humans: Studies	
Brooks et al.	(1946)	Nisbett	(1968a)
Teitelbaum & Campbell	(1958)	Nisbett & Gurwitz	(1970) [two experiments]
		Nisbett & Storms	(in press) [two experiments]
		Pliner	(Chapter 5, this volume)

TABLE 6
Speed of Eating

Animals		Humans	
Batting average	Mean O/N	Batting average	Mean O/N
1/1	1.28	1/1	1.26
Studies		Studies	
Teitelbaum & Campbell	(1958)	Nisbett	(1968b)

"any burst of food intake of at least five pellets separated by at least 5 min. from any other burst [p. 138]." For humans, these particular data are based on self-report or interview studies, for we know of no relevant behavioral data. Given this caveat, again the data for the lesioned rat and the obese human correspond closely.

From the previous two facts, it should, of course, follow that obese subjects will eat a fair amount more per meal than normal subjects, and as can be seen in Table 5, this is the case for both lesioned rats and obese humans. The data for rats are based on two experiments that simply recorded the amount of food eaten per eating burst. The data for humans are based on all experiments in which a plate of food, usually sandwiches, is placed before a subject, and he is told to help himself to lunch or dinner.

Our final datum on eating habits is the speed of eating. Teitelbaum and Campbell (1958) simply recorded the number of pellets their animals ate per minute. Since there is nothing else to do when you're sitting behind a one-way screen watching a subject eat, Nisbett (1968b—data not reported in paper) recorded the number of spoonfuls of ice cream his subjects ate per minute. The comparison of the two studies is drawn in Table 6, where there is an unsettling similarity in the rate at which lesioned rats and obese humans outspeed their normal counterparts.[6]

All told, then, in the existing literature we found a total of six items of behavior on which it is possible to make comparisons between lesioned rats and obese humans. These are mostly nonobvious facts, and the comparisons drawn between the two sets of experiments do not attempt to push the analogies beyond the point of common sense. For all six facts in the existing

[6]Fat rats do not drink a liquid diet more rapidly than do normals. There are no comparable data for humans.

literature, the parallels between the species are striking. What the lesioned, fat rat does, the obese human does.

In addition to these facts, we have identified several other areas of behavior, unrelated to food consumption, in which it is possible to draw somewhat more fanciful, though still not ridiculous, comparisons between the species. For example, on gross activity, studies using stabilimeter cages or activity wheels have demonstrated that the lesioned animal is markedly less active than the normal animal. This is not, it should be added, a totally trivial fact indicating only that the lesioned animal has trouble shlepping his immense bulk around the cage, for the dynamic, hyperphagic rat—who though not yet fat, will be—is quite as lethargic as his obese counterpart (Teitelbaum, 1957). On the human side, Chirico and Stunkard (1960) used a pedometer to measure the distance walked per day by obese and normal subjects who were matched for occupation and found that the obese were far less active than normals. Bullen, Reed, and Mayer (1964) report similar findings for degree of participation in scheduled athletic activities in a summer camp for girls.

In addition, the lesioned animal's emotionality and its apparent inability to adjust its intake to the caloric density of its diet both have suggestive, though far from conclusive, counterparts in the research literature on the obese human. Both of these areas of behavior will be considered in detail in other chapters.

So far, we have located no fact in the existing literature in which the parallel fails. There are, of course, still numerous facts about either the lesioned animal or the obese human for which parallel data do not, as yet, exist in the available research literature. To pursue this line of analogical investigation, we and our colleagues deliberately designed a program of research to pursue experimentally these parallels. In effect, we have designed a series of experiments for humans that have no particular rhyme or reason except that someone once tested such a relationship on a lesioned rat. The specific facts about the lesioned animal which we have attempted to test on the obese human are the following:

1. The VMH lesioned animal is a hyperemotional animal—easily startled, excitable, irascible, and difficult to handle.
2. The lesioned animal is better at active avoidance and worse at passive avoidance than is its normal counterpart.
3. The lesioned animal appears to be more sensitive to pain.
4. Though the lesioned animal appears unable to regulate its intake in accordance with the caloric density of its diet, this seems to be truer of animals tested on a solid rather than a liquid diet.
5. Though the lesioned animal will eat large quantities of easily-available food, it will not work to get food.

The following series of studies are experiments designed to test each of these relationships on human subjects.

2
THE EFFECTS OF WORK AND CUE PROMINENCE ON EATING BEHAVIOR

Stanley Schachter and Lucy N. Friedman

Among the most perverse of facts about the VMH-lesioned rat is the finding that though the lesioned animal will outeat its intact counterpart when food is freely available, it simply will not work to get food. Miller et al. (1950) first noted that though lesioned rats ate more than controls when an unweighted lid covered the food dish, they ate less when a 75-gram weight was fastened to the lid. They also found that the lesioned rats ran more slowly down an alley to food than controls did and pulled less hard when temporarily restrained by a harness. Extending this work, Teitelbaum (1957) studied the effects of various reinforcement schedules and demonstrated that at FR1, when one press yields one pellet, fat lesioned rats outpress normal. As the payoff decreases, however, fat rats press less and less until at FR256, they don't manage to get a single pellet during a 12-hour period, whereas normal rats are still industriously pressing away. Hamilton and Brobeck (1964) working with rhesus monkeys report that they do not replicate Teitelbaum's findings, but Mrosovsky's (1971) reanalysis of their data suggest that the lesioned monkey, too, will not work for food.

Though the fact seems reasonably clear, its interpretation is not, for virtually all manipulations of work have covaried the remoteness or prominence of the food cues. Food at the end of an alleyway is obviously a more remote cue than food in the animal's food dish. Pellets available only after 256 presses of a lever are certainly more remote food stimuli than pellets available after each press of a lever. Existing experimental designs, then, do not allow us to partial out the effects of cue prominence from those of the work required to get food.

The present experiment has two purposes: first, to determine if there is a human analogy to the Miller et al. (1950) and Teitelbaum (1957) findings on the hyperphagic rat, and second, to independently examine in humans the effects of cue prominence and of the effort required to get food.

Method

The experiment was conducted within the guise of a study of the interrelationship between personality and sociological variables which required of the subjects only that they complete a number of personality tests and sociological questionnaires. Each subject was scheduled to come to the experimenter's office some time between 1:30 and 5:00 in the afternoon. When he arrived, the experimenter was seated at her desk on which were papers, folders, a partially drunk cup of coffee, a nutcracker, and in a letter tray on the right side of the desk, a bag of almonds. As the subject sat down next to the desk, the experimenter ate a nut and explained the purported purpose of the study and the questionnaires. She then stood up and said, "I have to use the calculator for a few minutes so why don't you sit in my chair, I think you'll be more comfortable. You can use a pen or a pencil and you don't have to put your name on the questionnaires." She then took two more nuts and added, "There's no time limit, you can do it at your own pace. Help yourself to some nuts if you like." Then, while eating a nut, she closed the door behind her leaving the subject completely alone. After exactly 15 minutes, the experimenter returned, casually picked up her bag of nuts, sat down at the other desk in the same office, and busied herself scoring questionnaires.

When the subject had finished his job, the experimenter scored the personality tests, interpreted them for the subject, thanked him for coming, and when he had left, weighed the bag of nuts (which originally had weighed 464 grams) to learn if the subject had eaten and how much.

Manipulating Effort: In order to manipulate the effort necessary to get at food, there were two sets of experimental conditions. In one, the almonds had shells on them; in the other, they did not. To insure that taste or freshness was not a confounding factor, the almonds used in the shelled condition were always shelled shortly before each experimental session.

Manipulating Cue Prominence: To manipulate the conspicuousness of the food cue, in one set of conditions the nuts were in a transparent cellophane bag. In the other set of conditions, the nuts were in a brown, opaque paper bag.

Subjects: There were a total of 80 subjects, half normal, half obese, all male, Columbia undergraduates. Normals ranged from 12.6% underweight to 9.0% overweight; the obese from 15.2% to 63.7% overweight as calculated from the norms published by the Metropolitan Life Insurance Company (1959).

In summary, effort, cue prominence, and degree of obesity were covaried in an eight-condition experiment. There were 10 subjects in each condition.

Results

Table 1 presents the data in a form which makes it convenient to examine the effects of the work variable. The figures in the table are simply a frequency

TABLE 1

The Effects of Work on Eating Behavior

Nuts have:	Number of normal subjects who:		Number of obese subjects who:		χ^2	p
	Eat	Don't eat	Eat	Don't eat		
Shells	10	10	1	19	8.03	<.01
No shells	11	9	19	1	6.53	<.02
χ^2	0		28.90			
p	n.s.		<.001			

tally of the number of subjects who do and who do not eat nuts in those conditions where the nuts are shelled. It can be immediately seen that the work manipulation had almost no effect on normal subjects. When the nuts have shells, 50% of normal subjects eat; when the nuts have no shells, 55% eat. The obese stand in fascinating contrast. When the nuts have shells and eating requires the labor of using a nutcracker, only 1 of 20 fat subjects eat. When the nuts are shelled and eating is simply a hand-to-mouth operation, 19 of 20 fat subjects eat. There appears to be no question, within this experimental context, that the work involved has had a major effect on the likelihood that a fat person will eat. The Miller et al. (1950) and Teitelbaum (1957) findings on the lesioned rat appear to have a firm analogy in the obese human.

Turning next to the effects of cue prominence, Table 2 presents data comparing the effects of the cellophane versus the brown paper bag on eating behavior. It can be seen that at best this particular manipulation of cue prominence has had minimal effects; both normal and obese subjects are trivially and nonsignificantly more likely to eat when they can see the nuts than when they cannot.

Discussion

Certainly the clearest conclusion to draw from the results of this and conceptually equivalent experiments on lesioned animals is that labor, unconfounded by cue prominence, is a key variable in determining how much obese subjects eat. Unfortunately, the results of at least four other studies (Johnson, Chapter 8; Rodin, Chapter 13; Ross, Chapter 7; Goldman, Jaffa, & Schachter, 1968) make it likely that such a conclusion is wrong. Ross,

TABLE 2

The Effects of Cue Prominence on Eating Behavior

Bag is:	Number of normal subjects who:		Number of obese subjects who:		χ^2	p
	Eat	Don't eat	Eat	Don't eat		
Cellophane	12	8	11	9	0	n.s.
Brown paper	9	11	9	11	0	n.s.
χ^2	0.40		0.10			
p	n.s.		n.s.			

for example, finds that when a bowl of unshelled nuts is brightly illuminated, obese subjects eat far more than do normals. However, when the nuts are dimly illuminated the obese eat less than normals. The results of Johnson's experiment, in particular, challenge a conclusion that cue prominence is not an alternative explanation of the results of the experiments on labor and eating. As we did, Johnson independently manipulated cue prominence and effort. In contrast to our findings, his results indicate that obese subjects will work far harder than normal subjects when the food cue is prominent.

We suspect that there is a simple reason that our results differ from those obtained by these other investigators. Perhaps we have been hoist on the petard of an irresistibly cue manipulation of work. Nuts with shells on them are obviously harder to eat than nuts without shells, but they are also a more remote food cue than shell-less nuts. Whether the bag is cellophane or brown paper, the subject in the shell conditions cannot directly see the *edible* object. In short, the analogy to the animal experiments may be better than we intended, since both sets of experiments confound prominence and work. There appears to be no question that both the VMH-lesioned rat and the obese human will not work to get food when the food cue is remote. Whether labor or cue prominence is the determining variable, the reader can evaluate for himself when he has read the remaining studies in this monograph.

3
EMOTIONALITY AND OBESITY

Judith Rodin, Donald Elman, and Stanley Schachter

Other than its hyperphagia and associated eating peculiarities, probably the single most common observation made about the ventromedial-lesioned animal is that it is hyperemotional—easily startled, hyperexcitable, and generally difficult to handle. This observation has been made by virtually everyone who has worked with this preparation, and casual mention of this fact can be found in a large variety of experimental papers concerned with other facets of the VMH-animal's behavior (e.g., Brooks et al., 1946; Kling & Hutt, 1958; see also MacLean, 1969 for a review). Several investigators have systematically studied emotional reactivity on a variety of indices, such as aggression and irritability, and find that lesioned animals are considerably more emotional postoperatively than unoperated controls (Eichelman, 1971; Grossman, 1972; Paxinos & Bindra, 1972; Sclafani, 1971; and Singh, 1969). Grossman (1966), partially on the basis of his own work on avoidance learning, has proposed that emotionality can be considered the chief characteristic of the VMH animal and attempts to deduce the many peculiarities of its eating behavior from this fact. There appears to be no question that the lesioned animal is an emotional animal.

For humans, to date there has been no deliberate attempt to compare obese and normally-sized subjects on the kinds of behavior that could plausibly be interpreted as indicating emotionality. The single relevant datum in our past work comes from a study by Schachter, Goldman, and Gordon (1968)—their incidental and unpublished finding that obese subjects are more frightened by a threat of electric shock than are normal subjects. This experiment was, in part, concerned with the effects of fear on eating behavior. To manipulate fear, subjects were threatened with either very painful or very mild electric stimulation. To check on the success of the manipulation, subjects were asked to answer the following question:

How nervous or uneasy do you feel about taking part in this experiment and being shocked?

Extremely Uneasy (5)	Very Uneasy (4)	Quite Uneasy (3)	Slightly Uneasy (2)	Relatively Calm (1)	Very Calm (0)

In the high fear conditions, fat subjects averaged 2.20 on this scale and normal subjects 1.66 ($p = .07$). Other than this slight indication that the obese respond more fearfully in frightening situations, we know of no other experimental evidence relevant to the question of obesity and emotionality. To pursue this matter, two studies were conducted.

EXPERIMENT I. THE EFFECTS OF LISTENING TO EMOTIONALLY DISTURBING TAPES ON FEELINGS

Within the context of a larger study on the effects of distraction on performance, Rodin (Chapter 13) was able to examine the relative emotionality of obese and normal subjects by varying the nature of the distracting material, so that some subjects listened to intensely emotional recorded material and others to neutrally-toned unemotional material. The complete details of this study are reported in Chapter 13. For present purposes, it will suffice to note that in the conditions relevant to present concerns, each subject worked for a 10-minute period at either a proofreading task or a complex reaction-time task while listening to the tapes and also spent a 10-minute period during which he did nothing but listen to the tape. During this time, he listened through earphones to one of two kinds of tapes: (*a*) emotionally disturbing tapes, which for the 10 minutes detailed either the bombing of Hiroshima or the subject's own death via leukemia, or (*b*) emotionally neutral tapes, which were concerned with either rain or seashells. A few typical excerpts will allow the reader to judge the effectiveness of these tapes.

From Hiroshima:

". . . Picture how the eyebrows of some were burnt off, and skin hung from their faces and hands. Others were vomiting as they walked."

". . . Imagine trying to help someone, reaching down and grabbing him by the hands but not being able to hold him because his skin slipped off in huge, glove-like pieces into your hands."

From leukemia:

". . . Imagine how you would feel if you could feel your body degenerating—a constant deterioration going on inside you that no one was able to stop or even slow down. You would have to be fed because you would be too weak to hold a fork. You would have to lie in bed because you would be too weak to hold up your own head."

". . . Think about who in your family would help you, which of your friends would stand by you . . . . How would your illness affect their lives?

Who would feel inconvenienced and put upon? . . . Who would be glad to see you dying in great pain and suffering?"

From neutral tapes:

". . . Think about all the varied shapes and colors of shells you've seen. Think about seeing them along the beach, about picking them up and saving an unusual one. . . . Some are rough with spiny and irregular edges that tingle when you pick them up."

". . . Think about the rains that come in the spring and the fall. Remember how sometimes it rains so hard that the sewers plug up within minutes. You may be outside and before you can reach cover you are soaked and the ground is turned to mud."

Obviously, one set of tapes is concerned with deeply disturbing material and the other could be expected to leave the auditor untouched or, at best, in a lyrical, semi-abstracted mood.

Immediately after listening to the tape, all subjects answered a series of questions concerned with the task on which they had worked and with their perceived physiological and emotional state. The questions relevant to emotionality are the following:

1. Are you experiencing any palpitation?
2. Do you think your breathing rate is faster than usual?
3. Are you feeling generally upset?
4. Are you experiencing any anxiety?
5. Do you feel emotionally aroused?

Each of these questions was answered on an appropriately labeled version of the following scale.

0	10	20	30	40	50	60	70	80	90	100
not at all										extremely

The subjects in the study were all male, Columbia undergraduates—half obese, half of normal weight. The obese ranged from 15.5% to 62.8% overweight and normals from 9.8% underweight to 9.8% overweight.

Results

The effects of the tapes on subjects' self-report of feelings are presented in Table 1. The figures in this table represent an emotionality index derived simply by averaging each subject's answers to the five questions designed to measure emotionality. It is evident that obese subjects are considerably more disturbed by the emotional tapes than are normal subjects. That this difference does not simply reflect a penchant for the obese to check higher points on scales such as ours, is indicated by the results of the Neutral-tape con-

TABLE 1
Emotionality of Obese and Normal Subjects

Subjects	Mean Emotionality Index		p
	Neutral tapes	Emotional tapes	
Normal	16.6	20.1	n.s.
Obese	9.5	29.3	<.002
p	<.01	<.05	

ditions. When they listen to nondisturbing material, the obese describe themselves as significantly less emotional than do normal subjects.

The effects of these tapes on performance are described in detail in Chapter 13. To summarize, the obese do considerably and significantly worse than normals at either a proofreading or a reaction time-monitoring task when they are listening to the emotionally disturbing tapes. In those conditions that did not attempt to manipulate emotionality, the obese either do not differ from normals or do significantly better, depending on condition, at these tasks. By either self-report or performance measures, then, the obese appear to be considerably more affected by emotionally arousing stimuli.

EXPERIMENT II. THE EFFECTS OF PAIN ON LEARNING

Pain and the threat of pain are, of course, disturbing and unsettling experiences bound to make a subject edgy and tense. If it is correct that the obese are more reactive and emotional than are normals, it should be expected that the experience and anticipation of pain will be more disruptive for obese than for normal subjects. In part to test this hypothesis, an experiment was designed to test the effects of pain on the ability of obese and normal subjects to learn a rather complex mental maze.

To measure learning we employed Lykken's (1957) electronic maze, which is a 20-step maze with 4 alternatives at each choice point. The apparatus proper was housed in a small metal box with four levers mounted in the front. The subject's job was to thread his way through the maze by pressing the correct sequence of levers. At each step, when a subject pressed the correct lever, a green light flashed on and the machine automatically moved on to the next step. If the subject pressed one of the three incorrect levers, a red light flashed and an error was recorded on a counter visible to the subject. A light signalled when the subject had worked his way through the maze, the machine was reset, and the subject began working again until he reached a criterion of three errorless trials or had completed 21 trials.

There were three experimental conditions—high shock, low shock, or no shock. In the shock conditions, electrodes were fastened to the second and third fingers of the subject's nondominant hand. The experiment had been explained as a study of the effects of "partial reward and punishment" on learning and the subject was told, ". . . you will occasionally receive a shock when you make an error. We have set up the apparatus to randomly give a shock once every few errors." In fact, the shocks were not administered randomly but were linked to one of the three incorrect levers at each point. Theoretically, a subject could learn to avoid shock by not pressing the shock lever at a given step, even if he had not learned the correct lever. In the no-shock condition, since we wanted to minimize the possibility of upsetting the subject, electrodes were not fastened to a subject's hand and no mention was made of shock.

In order to find the appropriate shock level for each individual subject in the shock conditions, the experimenter told the subject:

> Before we begin the learning task, I must test your sensitivity to electric shocks. I will present a series of brief shocks, one every 10 seconds, beginning from an imperceptible level and gradually increasing in intensity.—I would like you to tell me when you first feel a noticeable tingling sensation.

The first shock, delivered by a constant-current shock generator, was .06 ma. and each succeeding one was increased by .01 ma. Shock duration was .5 seconds. Shock presentation continued until a subject made two successive positive responses. This defined the sensation threshold. To determine pain threshold, the experimenter continued presenting shocks in increasing steps of .05 ma. The pain threshold was the level at which the subject reported first feeling "definitely painful."

In the high-shock condition the shock level for each subject was set at .25 ma. above his pain threshold. In this condition, to compensate for adaptation as the experiment progressed, shock intensity was increased by .05 ma. each time a subject received 10 shocks.

In the low-shock condition, the shock level was set at a point .25 ma. below the pain threshold. Since in a few cases this would have pushed a subject below his sensation threshold, we adopted the arbitrary rule of never setting the low-shock level lower than .10 ma. above the sensation threshold.

There were a total of 90 subjects, all male, Columbia undergraduates, half obese, half normal, and equally divided among the experimental conditions. Normals ranged from 8.6% underweight to 10.0% overweight. The obese ranged from 15.1% to 51.3% overweight.

TABLE 2
Effects of Electric Shock on Learning

	Total number of errors made in:			
Subjects	No shock	Low shock	High shock	No vs. high-shock *p*
Normal	228.7	163.4	198.1	n.s.
Obese	189.9	228.8	286.5	<.05
p	n.s.	n.s.	<.05	
	Interaction $p < .03$			

Results

The effects of electric shock on learning are presented in Table 2. The figures in the table are the average of the total number of errors, shocked and unshocked, made by each subject in the course of learning the maze. It is evident that shock has an increasingly disruptive effect on the learning ability of obese subjects and no such effect on normal subjects. The obese make somewhat fewer errors in the no-shock condition than do normals and considerably more errors in the high-shock condition (interaction $p < .03$). It does appear that, compared with normal subjects, pain interferes with the ability of the obese to learn a complex task.

Altogether, then, there are three experimentally garnered facts which support a view of the obese as more emotional than normals.

1. In response to a threat of painful shock, the obese describe themselves as more nervous than do normals.
2. Emotionally distressing audio tapes are more upsetting to obese than to normal subjects.
3. Painful shock interferes with the ability of the obese to learn a complex task.

It does appear reasonable to conclude that the emotionality of the VMH-lesioned animal is paralleled by the emotionality of the obese human.

4
SHOCK AVOIDANCE BEHAVIOR IN OBESE AND NORMAL SUBJECTS

Judith Rodin

Among the behaviors which distinguish the ventromedial-lesioned animal from its intact counterpart is its performance in shock-avoidance situations. Typically, the lesioned animal learns a response to avoid shock far more quickly than does a normal control. In these active avoidance studies, subjects are usually presented with a conditioned stimulus, such as a light or buzzer, which precedes shock or other noxious stimulation. To avoid shock, they must actively make some response, such as pressing a lever or crossing a barrier within a specified period of time. For example, Sechzer, Turner, and Liebelt (1966), using goldthioglucose-lesioned mice, trained animals to avoid shock by running down a maze within 5 seconds. They found that lesioned mice take fewer than half the number of trials needed by nonlesioned controls to reach criterion. Similarly, Grossman in two studies (1966, 1972), Levine and Soliday (1960) and Sepinwall (1969) also report that VMH-lesioned animals are superior to normals in active avoidance learning. On the other hand, two studies (Grossman, 1970; McAdam & Kaelber, 1966) find that lesioned animals learn an active avoidance response more poorly than do normals.

Since, in this area, a variety of lesion-producing techniques have been used, including electrolytic, goldthioglucose, topical application of atropine, and knife cuts, the use of a batting-average approach here is undoubtedly somewhat questionable, since the studies may not be strictly comparable. Nonetheless, in five of seven experiments, the lesioned animals do better than controls. In a maze or shuttelbox situation they learn more rapidly; in a Sidman avoidance procedure, they work harder to avoid shock. In short, it appears that lesioned animals will work harder and more efficiently to avoid a painful experience than will normal animals in a simple avoidance situation.[1] If there is a human parallel to this active avoidance situation, it should be expected that obese humans will also devote more effort to avoiding shock.

Method

An experimental situation was devised to examine the extent to which a human subject tries to avoid shock in a simple avoidance task. Within the context of a study of problem solving, subjects were told that they would receive a shock unless they solved a particular puzzle within a 3-minute period. They were given the choice of working either on this puzzle or on a puzzle whose solution would yield a small monetary reward.[2] They could work on either or both puzzles, but unless they solved the shock-avoidance puzzle they were certain to be shocked. We assume that the proportion of time that they spent working on the shock-avoidance puzzle is a loose analog to the effort the experimental animal exercises to avoid shock.

To make the threatened shock vivid, the experimenter put on a white lab coat and effortfully wheeled a large shock apparatus into view. Electrodes were attached to each subject's wrists with a liberal smattering of electrode paste, and the shock apparatus was turned on and apparently calibrated.

The subjects then listened to tape-recorded instructions about the task. They were told that they would receive two envelopes, each containing four pieces which fit together to form an 8 × 10 rectangle. By solving the red puzzle, they could avoid shock, and by solving only the green puzzle, they would receive money but would be shocked. The amount of money and intensity of shock were left unspecified. They were told they had a total of 3 minutes and, during that time, the order in which they worked on the puzzles and the amount of time spent on each was entirely up to them. Subjects then received two envelopes, one marked "Shock" and the other "Money." Actually both puzzles were insoluble, although pretesting had established that this was not apparent. At 10-second intervals, the experimenter recorded on which of the puzzles each subject was working.

The subjects in the study were all female, New York University undergraduates—half obese and half of normal weight. Obese subjects ranged from 15.3% to 38.8% overweight and normals from −7.8% to +6.5% over-

[1]For humans, the data also suggest that when the learning task is quite complex, such as the mental maze described in the preceding chapter, shock-produced arousal may interfere with learning for the obese. While we know of no comparable complex avoidance study using lesioned animals, we might expect a similar effect: The heightened responsiveness of the obese to shock (Turner, Sechzer, & Liebelt, 1967) should increase arousal relative to normals. As Zajonc (1965) has demonstrated, arousal facilitates performance on simple tasks; however, as the learning situation becomes more complex, increased arousal disrupts performance.

[2]The experimental task was modified from Ross, Rodin, and Zimbardo (1969).

TABLE 1
Mean Time (in seconds) Spent Working on Each Puzzle

Subjects	Shock avoidance[a]	Reward
Normals	127.69	52.31
Obese	162.31	17.69

[a] p value = .05

weight. Three or four obese and normal subjects were tested at a time, and they sat in 3′ × 3′ cubicles separated from one another.

Results and Discussion

If there is a parallel to the behavior of VMH-lesioned animals in simple, active avoidance, obese subjects should spend more time working on the shock-avoidance puzzle. Of 13 obese subjects, 8 spent the entire time allotted to the puzzles working exclusively to avoid shock. In contrast, only 4 normals spent the full 180-second period on the shock-avoidance puzzle. Overall, fat subjects spent a significantly greater amount of time than did normals working to avoid shock ($p = .05$ two tailed[3]). The mean number of seconds spent on each puzzle is given in Table 1.

Subjects were told that they could work on the puzzles in any order they wished. Thus, seeing the puzzle on which they began should reflect the relative priority they assigned to solving each one. As indicated in Table 2, all 13 overweight subjects began with the shock-avoidance puzzle, whereas 5 of the 13 normals started by working to obtain the reward. This difference, using the Fisher exact-probability test, is significant with $p = .05$.

These data indicate that overweight subjects were, in fact, making a greater effort than normals to avoid the threatened shock. More of the obese began working on the shock-avoidance puzzle, and they used more of the available time trying to solve it than did normals. Although we did not put human subjects through an actual avoidance-learning procedure, in a broad sense a parallel does appear to exist between obese humans and lesioned animals. When threatened with electric shock, both kinds of subjects work

[3]In order to correct for the large number of tied scores, the normal approximation to the sampling distribution of U, corrected for ties, was used (Siegel, 1956, p. 231).

TABLE 2

Number of Subjects Who Began Working on Each Puzzle [a]

Subjects	Shock avoidance	Reward
Normals	8	5
Obese	13	0

[a] *p* value = .05

more efficiently, harder, or longer than do normal controls to avoid painful stimulation.

5

EFFECTS OF LIQUID AND SOLID PRELOADS ON THE EATING BEHAVIOR OF OBESE AND NORMAL PERSONS[1]

Patricia Pliner

The hypothesis that the human obese do not regulate food intake as well as normals has been supported by a variety of studies which indicate that the amounts eaten by the obese and their self-reports of hunger do not vary with food deprivation or with the physiological symptoms that we associate with deprivation. Unlike normals, the subjective report of hunger in the obese appears to be relatively independent of food deprivation (Nisbett, 1968b) and of gastric motility (Stunkard & Koch, 1964). Similarly the obese are unresponsive to experimentally manipulated deprivation. Schachter et al. (1968) fed roast-beef sandwiches to hungry subjects in one condition and did not feed another group of subjects in a comparison condition. In subsequent measures of eating, normal subjects adjusted their intake in accordance with preloading, but obese subjects ate as much when they had eaten roast-beef sandwiches as when they had been deprived. In addition, several studies of fasting behavior (Goldman et al., 1968; Schachter & Gross, 1968) support the general conclusion that the obese are relatively insensitive to the physiological correlates of food deprivation.

There are, however, at least two studies which do not support this conclusion. Nisbett and Storms (in press) preloaded underweight, normal, and obese subjects with 25 ounces of Nutrament, containing 750 calories, or 25 ounces of Diet Pepsi, containing two calories. Two hours later they presented

[1]This article is based on a dissertation submitted to the Department of Social Psychology of Columbia University in partial fulfillment of the requirements for the Ph. D. degree. The author is greatly indebted to Stanley Schachter for stimulation and guidance in all phases of the work. Howard Cappell, John Arrowood, and J. Barnard Gilmore contributed valuable criticism.

subjects with a test meal. Consistent with expectation, normal and underweight subjects responded strongly to the caloric manipulation, eating less following the Nutrament preload than following the Diet Pepsi preload; but contrary to expectation, obese subjects also responded strongly to the caloric manipulation. In a similar study, Nisbett and Storms (in press) preloaded subjects with 25 ounces of Nutrament or a noncaloric pill and presented a test meal five minutes later. Again obese subjects, as well as normal and underweight subjects, responded to the caloric manipulation.

While there were many procedural differences between the Schachter et al. (1968) study and the Nisbett and Storms studies, perhaps the most obvious was that the preloads were solid foods in the former and liquid foods in the latter. It is the purpose of this study to explore possible differential effects of liquid and solid preloads on obese and normal human subjects.

It should be noted immediately that this experiment was *not* conceived originally to test a parallel between obese humans and hyperphagic rats. However, consideration of the research literature suggests that the liquid-solid distinction may also clarify inconsistent results obtained with these animals. Much of the available evidence on the VMH-lesioned, hyperphagic rat indicates that it, unlike its normal counterpart and like its obese human counterpart, is unresponsive to internal cues. On the one hand, the hyperphagic rat's gross overeating can be taken as evidence for its unresponsiveness to internal satiety cues, and on the other, its failure to compensate for caloric dilution of its chow (Kennedy, 1950; Teitelbaum, 1955) and for extended enforced fasts (Miller et al., 1950) can be taken as evidence for its lack of responsiveness to internal deprivation cues.

However, for hyperphagic rats as well as for obese humans, there are exceptions to the finding of internal unresponsiveness. Compensation for caloric dilution of liquid diets by both normal and hyperphagic rats was found in a study by Williams and Teitelbaum (1959). Smith, Salisbury, and Weinberg (1961) preloaded normal and hyperphagic rats by stomach tube with water, a glucose solution, or a saline solution and then allowed them to eat *ad lib* for two hours. Both groups of rats displayed internal responsiveness, eating less following glucose preloads than following water or saline preloads. Thomas and Mayer (1968) report that hyperphagic rats as well as control rats responded appropriately to continuous gastric infusion of both liquid diet and water, decreasing oral intake when infused with liquid diet but not when infused with water.

Thus from the available evidence, it is possible that still another parallel between the behavior of VMH-lesioned rats and obese humans exists. Both types of obese organism appear to be differentially responsive to internal cues arising from solid and liquid foods. Evidence of differential responsiveness from these two independent areas of research lends some convergent validity to the notion that a distinction between solids and liquids is a potentially fruitful one. Of course, such a distinction is at this point merely an empirical generalization and is virtually devoid of conceptual significance.

The present study was designed to test the differential-responsiveness hypothesis for human subjects. The amount of food eaten by obese and normal subjects was measured at a test meal following ingestion of four different preloads: high-calorie solid, low-calorie solid, high-calorie liquid, and low-calorie liquid. The predictions are straightforward. If the hypothesis that the obese are differentially responsive to solids and liquids is correct, the caloric manipulation should affect the eating behavior of obese subjects given liquid preloads, while the eating behavior of obese subjects given solid preloads should remain unaffected. Normal subjects should, of course, respond to both caloric manipulations.

METHOD

Procedure

Subjects were recruited for an experiment, the ostensible purpose of which was to measure the effect of a vitamin on ability to concentrate. At the time of recruitment, subjects were asked not to eat on the day of their experimental appointments and as a reward for fasting were promised lunch at the end of the experiment. Upon his arrival at the laboratory, each subject was told that he would be given a capsule containing either the vitamin or a placebo, and would then be asked to perform a "concentration task." The necessity for fasting was explained as a means of ensuring that subjects did not consume any of the vitamin in its natural form with the food they ate.

Subjects, half normal, half obese, were scheduled for one of three times of day: noon, 1:45 p.m., or 3:30 p.m. Four subjects in each condition were run at each of the three time periods. Since preliminary analyses showed no effect of the time variable, data for the three time periods were combined. Whenever possible, two subjects were scheduled for each experimental session. The pairs of subjects received the experimental instructions in the same room, were placed in different rooms to work on the "concentration task," and remained separated until the end of the experiment. Sixty-eight of the 96 subjects were run in pairs, and the remaining 28 subjects were run singly. Since examination of the data revealed no systematic effect of this slight difference in procedure, the two groups of subjects were pooled for all analyses.

Preloads

After hearing the introductory spiel and before taking the capsule, each subject was asked to ingest one of the four preloads under the pretext that it was necessary to ensure proper absorption of the vitamin.

Liquid Preloads. The 600-calorie preload consisted of 400 cc. of a formula prepared with a skim milk base to which was added butter, sodium caseinate, dextrose, and water. The 200-calorie preload consisted of 400 cc. of a one-third dilution of the above formula, thickened with carageenan, a noncaloric substance. Both preloads were flavored with a noncaloric strawberry flavoring and were served in tall glasses resembling those used in soda fountains. The two preloads were virtually identical in appearance.

Solid Preloads. The 600-calorie preload consisted of a medium-sized piece of white cake, sliced into layers filled with strawberry jelly and covered with white icing. The 200-calorie preload consisted of a medium-sized piece of angel-food cake, sliced into layers filled with dietetic strawberry jelly and covered with white icing. The cakes were served on paper plates, and subjects were provided with dessert forks. The two preloads were virtually identical in appearance.

Caloric Composition of Preloads. Nutritionally, the solid and liquid preloads differed from one another in one obvious respect. The compositional breakdown of the liquid preloads was approximately 45% carbohydrate, 15% protein, and 40% fat. For the solid preloads the breakdown was approximately 84% carbohydrate, 6% protein, and 10% fat. While it would have been desirable to use solid and liquid preloads with identical caloric compositions, preliminary work proved this to be unfeasible.

Test Meal

Following ingestion of the preload, each subject was given a capsule allegedly containing "either the vitamin or the inert substance," but in fact containing a placebo in every case. Scheduled next was the concentration task, intended merely to be consistent with the rationale of the experiment and to allow a one-hour interval between ingestion of the preload and presentation of the test meal. Following the one-hour interval, each subject was presented with a test meal and told, "Here is the lunch I promised you—it's chicken and roast-beef sandwiches. Have all you want because I have lots more. I'll be back in a little while."

Subjects were given twelve minutes in which to eat the test meal, which consisted of three roast-beef sandwiches and three chicken sandwiches cut in quarters. Each quarter-sandwich contained about 60 calories, divided roughly equally between bread and meat filling.

Measurement

Preload Ratings. Within the solid and liquid conditions, an attempt was made to equate the 200- and 600-calorie preloads for taste, appearance, and texture, in order that subjects would have identical cognitions concerning the

TABLE 1
Quarters of Sandwiches Eaten

Calories	Normal			Obese		
	Solid	Liquid	Solid and liquid	Solid	Liquid	Solid and liquid
200	8.17	7.79	7.98	7.29	8.08	7.69
600	4.92	4.10	4.51	7.17	6.00	6.59
200 and 600	6.55	5.95	6.25	7.23	7.04	7.14

caloric value of the preloads. To measure the success of this attempt, each subject was asked immediately after the test meal to rate his preload on 7-point scales on two dimensions—richness and heaviness.[2] The data showed that subjects did rate the 600-calorie preloads as richer and heavier than the 200-calorie preloads; however, there were no significant interaction effects involving weight, and correlations between ratings of the preloads and number of sandwiches eaten at the test meal were low and nonsignificant. Therefore, it seems unlikely that subjects' cognitions about the caloric value of the preloads influenced the outcome of the experiment.

Subjects

The subjects in the experiment were 48 obese and 48 normal undergraduate males attending Columbia University. As in all other studies in this series, the criterion for obesity was a 15% deviation from the average weight, corrected for height, presented in the Metropolitan Life Insurance Tables (1959); and, a subject was considered normal if his weight was 10% or less above this corrected average. Normal subjects ranged from –5.6% to +9.3% overweight with a mean of +1.2%. Obese subjects ranged from +14.7% to +63.7% with a mean of +27.3%.

RESULTS

The effects of liquid and solid preloads are presented in Table 1, which shows the mean number of sandwich-quarters eaten by subjects in each con-

[2]Informal pretests indicated that high ratings on these two dimensions connoted high-calorie content while low ratings connoted low-calorie content.

TABLE 2
Analysis of Variance for Quarters of Sandwiches Eaten

Source	*SS*	*df*	*MS*	*F*	*p*
Weight (*W*)	19.08	1	19.08	3.65	n.s.
State (*S*)	3.68	1	3.68	< 1	n.s.
Calories (*C*)	125.58	1	125.58	24.06	< .01
$W \times S$	1.00	1	1.00	< 1	n.s.
$W \times C$	33.61	1	33.61	6.45	< .05
$S \times C$	8.65	1	8.65	1.65	n.s.
$W \times S \times C$	3.44	1	3.44	< 1	n.s.
Error	459.45	88	5.22	—	—

dition. The analysis of variance in Table 2 shows a significant main effect of calories. This is hardly surprising since in every pair of conditions subjects given the 200-calorie preloads ate more than their counterparts given the 600-calorie preloads. However, this effect was accounted for largely by the behavior of normal subjects. Treating the weight groups separately, there was no significant main effect of calories for the obese group.

It can also be seen that the interaction between weight and calories is significant, replicating Schachter's (1968, 1971a) earlier findings that the obese are generally less sensitive to internal cues than are normal subjects. The caloric manipulation had a much greater effect on normal subjects than on obese subjects. Normal subjects given the low-calorie preloads ate 7.98 sandwich-quarters, while normal subjects given the high-calorie preload ate only 4.51 sandwich-quarters. Obese subjects in the low-calorie conditions ate 7.69 sandwich-quarters, while obese subjects in the high-calorie conditions ate almost as much—6.59 sandwich-quarters.

The analysis-of-variance effect most relevant to the differential-sensitivity hypothesis is the second-order interaction (weight × state × calories). This interaction is not significant, suggesting that the obese may not, in fact, be differentially sensitive to solid and liquid foods. However, closer examination of the means in Table 1 suggests that abandoning the differential-sensitivity hypothesis would be premature. Normal subjects given a 200-calorie solid preload ate 3.25 quarter-sandwiches more than normal subjects given a 600-calorie preload, while obese subjects showed a trivial corresponding difference of only 0.12 quarter-sandwiches. Comparison of the relevant means for subjects given the liquid preloads yields a difference of 3.69 quarter-sandwiches for normals and an appreciable difference of 2.08 quarter-

TABLE 3
Planned Comparisons of Mean Number of Quarter-Sandwiches Eaten

	Normal solid	Normal liquid	Obese solid	Obese liquid
Difference	3.25	3.69	0.12	2.08
F	12.14	15.66	0.02	4.97
p	$< .01$	$< .01$	n.s.	$< .05$

sandwiches for the obese. Planned comparisons between the pairs of means are presented in Table 3.

As expected, both comparisons were significant for normal subjects, who ate less following 600-calorie preloads than following 200-calorie preloads in both the solid and liquid conditions. Obese subjects failed to respond to the caloric manipulation in the solid conditions. In contrast, however, obese subjects in the liquid conditions ate significantly fewer quarter-sandwiches following the 600-calorie preload than following the 200-calorie preload.

For the purpose of demonstrating differential sensitivity it would, of course, have been more comforting to find a significant second-order interaction in the analysis of variance; however, in light of the fact that only one of many possible interactions was of interest, the failure to find a significant second-order interaction is not surprising. But the combination of an F of 4.97 in the obese liquid group, and the reassuringly low F in the obese solid group strongly suggests that the obese are indeed differentially responsive to internal cues arising from solid and liquid foods.

One can look at the data in a still more precise manner by testing only the one interaction which is of interest. The differential-responsiveness hypothesis predicts that the difference in amounts eaten following the 200- and 600-calorie preloads should be smaller for subjects in the obese solid group than for subjects in the obese liquid, normal solid, and normal liquid groups; a linear comparison of means shows that this prediction is strongly confirmed ($F = 7.18$, $p < .01$). Thus, a precise test of the one interaction which is relevant to the differential-responsiveness hypothesis is highly significant.[3]

[3]This comparison is not, of course, statistically independent of the planned comparisons presented in Table 3. It should, therefore, be considered an alternative, rather than a supplement, to the analysis in Table 3.

To summarize, the differential-responsiveness prediction was confirmed. Obese subjects adjusted their caloric intake following liquid preloads and failed to adjust their caloric intake following solid preloads; normal subjects adjusted their caloric intake with both solid and liquid preloads. In addition, the obese were found to be less sensitive overall to internal cues than normals, a replication of earlier findings.

DISCUSSION

The present experiment replicated within one study the results of Schachter et al. (1968), who found that obese subjects unlike normals were unresponsive to internal cues following solid preloads, and the results of Nisbett and Storms (in press), who found that obese subjects did respond to internal cues following liquid preloads. Thus, at the level of empirical generalization there seems little doubt that obese human subjects respond differentially to the internal cues consequent to liquid and solid preloads.[4] Thus, it appears that still another parallel exists between the behavior of obese humans and that of hypothalamic, hyperphagic rats. But, of course, the question remains: Why are the obese responsive to internal cues in the case of liquids, yet unresponsive with solids, while normals are responsive to both types of preload? There would appear to be at least two plausible explanations for this difference. First of all, it is possible that the cues signalling nutritional status provided by ingestion of liquid foods are different from those provided by ingestion of solid foods, and the obese differ qualitatively from normals, being responsive to the first set of cues yet unresponsive to the second. On the other hand, it is possible that liquids and solids provide the same physiological cues, but that the cues provided by

[4]Several recent studies by S.C. Wooley and O.W. Wooley are also relevant here. O.W. Wooley (1971) found that, while compensation was by no means perfect, both obese and normal subjects maintained on a liquid diet for 15 days responded to a change in the caloric density of their food. Subjects in both weight groups consumed a significantly smaller volume of a high-calorie version of the diet than a version containing approximately half as many calories; however, the reduction in volume consumed did not compensate completely for the difference in caloric content. Wooley, Wooley, and Dunham (1972) found that at several different time intervals after a liquid meal subjects in both weight groups correctly guessed significantly more often than chance whether they had been given a high- or low-calorie meal; however, the level of correct guessing was not impressive (e.g., one hour after the meal, 57% for obese subjects and 59% for normal subjects). In addition, both obese and normal subjects reported greater hunger following low-calorie meals than following high-calorie meals at every measured time interval after the meal. Thus, two additional studies indicate that when liquid foods are used, obese subjects appear to be as responsive to internal cues as normal subjects.

Finally, data from a study by S.C. Wooley (1972) indicate that neither obese nor normal subjects adjusted their short-term intake in response to a manipulation of the caloric content of a liquid preload. However, the study also included a strong manipulation of the *apparent* caloric content of the preloads, and it is likely that this manipulation and associated demand characteristics overshadowed any evidence of responsiveness to the real caloric content of the preloads.

liquids are stronger or more easily discernible than those provided by solids, and obese subjects differ only quantitatively from normals, being generally less responsive to internal cues and failing to adjust their intake only when the cues are not easily discernible.

The data contain a hint that the latter possibility may be true. Examination of the sandwich data for normal subjects shows that the difference between the number of sandwich-quarters eaten in the high- and low-calorie solid conditions was 3.25, while the corresponding difference in the liquid condition was 3.69. This difference of differences is not statistically significant, but it does show a slight tendency on the part of normals to respond more strongly to the liquid manipulation than to the solid manipulation and suggests that liquids may indeed provide more easily discernible cues than solids as to nutritional status. If this is the case, then the difference between obese and normal subjects may well be quantitative rather than qualitative. However, since the evidence is so tenuous, it may be best to consider this question unresolved.

Returning to the question of a parallel between the obese human and the lesioned rat, it does seem to us that as far as the liquid-solid distinction goes, the available evidence is strongly suggestive but not conclusive. The chief obstacle to a firm conclusion is that the phenomenon has not yet been adequately demonstrated for animals. The inference that differential responsiveness exists in hypothalamic, hyperphagic rats is based on a composite of the results of separate studies. In no single study was the solid-liquid variable systematically manipulated; rather those experiments which used solid food (e.g., Kennedy, 1950) demonstrated failure to regulate, while those which used liquid food (e.g., Smith et al., 1961) demonstrated successful regulation. Since these studies were done by different individuals and differed procedurally, it is clear that variables other than the liquid-sold one could account for the differential results.

There is, however, one suggestive case in which the same experimenter has looked at regulation in hypothalamic, hyperphagic rats with both solid and liquid foods. Strominger, Brobeck, and Cort (1953) found that hyperphagic rats unlike normals did not respond to the dilution of solid food with Ruffex. Almost incidentally they also reported that "when a diet of ground chow pellets diluted with an equal part of water (1:1) was further diluted with two parts of water (1:2) and then with three parts of water (1:3), the calculated dry food [and] . . . caloric . . . intakes were unchanged for two obese rats selected from a large group and for their two normal controls [p. 69]." Unfortunately, because only four animals were used and because no further procedural details were given, this report provides only anecdotal evidence for differential regulation.

A further obstacle to a firm conclusion of a parallel is the fact that the studies of regulation in rats have generally employed a different experimental paradigm than that employed in our human studies. Typically, the animal experiment compares average daily caloric intake of a control diet with the intake of a diluted or concentrated form of the same diet—a procedure which probably covaries taste with caloric value—whereas the human studies vary the caloric value of a preload and measure subsequent intake of a standard diet.

Recognizing these impediments to drawing a firm parallel we simply note these similarities between the animal and human data: (*a*) When preloaded with solid food of manipulated caloric value, normally-sized humans regulate subsequent food intake and the obese do not. When fed solid diets of varying caloric value, normal animals regulate intake and VMH-lesioned animals do not. (*b*) When preloaded with liquid food, both normal and obese humans regulate. When fed liquid diets, there are indications that both control and lesioned animals also regulate intake.

6
PAIN SENSITIVITY AND PASSIVE AVOIDANCE

The studies reported so far have consistently supported the assumption of a parallel between the behavior of the obese human and the ventromedial-lesioned animal. Experiments concerned with emotionality, active avoidance, the regulation of food intake, and with the effects of work have all indicated that, in a broad sense, obese humans and lesioned animals behave in similar fashions. In addition to these experiments, we have attempted to study pain sensitivity and passive-avoidance behavior. In the animal literature, lesioned animals are reported to manifest greater sensitivity to electric shock and are also reported to do worse at passive avoidance than do normal controls. Our attempts to study these relationships in humans have been unsatisfactory. The results on pain sensitivity are equivocal, with one study supporting the parallel and another indicating no difference between obese and normal humans. Similarly, in a single attempt to test passive avoidance we found no differences between our experimental groups, but we believe that our experiment was an inadequate test. In short, we are simply unsure as to how to interpret these results.

We must concede considerable discomfort about our uncertainty regarding these studies, for by raising issues such as the adequacy of an experimental design, we place ourselves in the extraordinarily uncomfortable position of unhesitatingly accepting the results of experiments which support a parallel and questioning only those studies which fail to demonstrate a parallel. Because of this we feel it is important to present the results of experiments which, under other circumstances, might better have waited on publication until further research had clarified the status of a relationship. In this way the reader can best evaluate for himself the entire body of current data on which rests the argument for a parallel between the obese human and the lesioned animal. For this reason too, we feel it particularly important to note that we have reported in this volume *everything* that we have done—in the library and in the laboratory—to explore this parallel.

PAIN SENSITIVITY

Eichelman (1971), Sechzer et al. (1966), and Turner et al. (1967) have demonstrated that lesioned animals are considerably more sensitive to electric shock than are their normal controls. In their test procedure, Turner et al. simply placed their animals in a wooden box with a metal grid floor and electrified the grid with shocks of 1-second duration that varied in intensity from 0.05 ma. to 0.25 ma. They observed flinch and jerk thresholds—a flinch being defined as "a sudden, startled movement" and a jerk as "a violent and sudden movement of the body [p. 237]." On both measures, lesioned rats and mice prove to have lower thresholds than do normals.

Two attempts were made to test this relationship on human subjects. In chronologically the first of these, a measure of electrical sensitivity was included as one of the battery of tests of general stimulus sensitivity described by Rodin, Herman, and Schachter (Chapter 12). Electrodes were taped to the subject's upper arm and shocks of approximately 0.1-second duration were delivered at roughly 10-second intervals by a constant-current shock generator. Starting at a subthreshold level, the shocks were gradually increased in intensity. To begin, the subject was asked to report when he first felt "any sensation" in the vicinity of the electrodes. Using the method of ascending and descending limits, a sensitivity threshold was established. The experimenter then continued delivering shocks and asked the subject to report "the first time you experience pain, that is, the first time it hurts." This point is designated the "pain" threshold. Once established, the experimenter said, "Now we'd like you to take a few more shocks. From now on you can stop whenever you like. Just tell us when the shocks are painful enough so that you'd rather not continue"—a point designated as the "endurance" threshold.

The results of this test of electrical sensitivity are presented in Table 1 where it can be seen that there is a marked difference between obese and normal subjects on sensitivity thresholds and a suggestive, though nonsignificant, trend on pain thresholds.

In order to retest this relationship in a different experimental context and with different apparatus, the procedure described earlier for taking pain and sensitivity thresholds was introduced into the study of the effects of pain on learning (Chapter 3). The mean thresholds for the two groups of subjects are presented in Table 2 where it can be seen that there are simply no differences between obese and normal subjects.

Obviously these two tests of the relationship have yielded incompatible results. Since we know of no compelling reason for assuming that one set of data is any closer to the truth than the other, we would suggest that the most conservative conclusion is that there is no difference in shock sensitivity between obese and normal humans and that the relationship established for rodents has not been demonstrated for humans. Our sole caveat is the

TABLE 1
Electric Shock Sensitivity of Obese and Normal Subjects

Subjects	N	Sensitivity threshold (ma.)	Pain threshold (ma.)	Endurance threshold (ma.)
Obese	20	.12	.66	1.56
Normal	20	.23	.95	1.58
t		3.05	1.48	< 1
p		< .01	< .20	n.s.

possibility that the human experiments are not an adequate parallel to the animal studies. Of necessity, the animal experimenters employed behavioral criteria such as jumping and flinching as indicators of shock sensitivity. Though certainly it would have been possible to observe counterpart behaviors for humans, it simply did not occur to us to do so, and we relied solely on verbal reports. Conceivably, sensory thresholds could be identical for obese and normal subjects, yet, as is the case for animals, obese humans might still prove to be more reactive than normals in the sense of flinching, pulling away, and so on.

PASSIVE AVOIDANCE

In passive avoidance, animals must learn not to respond in order to avoid shock. Typically, the animal receives a shock at the food or water dish if he eats or drinks during the experimental period. Three studies of passive avoidance (Kaada, Rasmussen, & Kveim, 1962; Margules & Stein, 1969; Sclafani and Grossman, 1971) found that lesioned animals showed poor avoidance learning, receiving a greater number of shocks than did normals. Sclafani and Grossman (1971) also reported one test in which there were no significant differences between groups. In all but one study, then, lesioned animals do more poorly at passive avoidance learning than do normals.

To test whether or not there is a human parallel for this behavior, an attempt was made to develop a conceptually equivalent passive-avoidance situation for human subjects. Employing precisely the same design as in the experiment on active avoidance (Chapter 4) a different group of subjects was given the option of working on either of two puzzles. In this case, however, the solution of one puzzle would win the subject $5.00, but she would be shocked; solving the other puzzle would earn her $.25, and she would not be shocked. There were no differences of any consequence between obese and

TABLE 2
Electric Shock Sensitivity of Obese and Normal Subjects

Subjects	*N*	Sensitivity threshold (ma.)	Pain threshold (ma.)
Obese	30	.15	.65
Normal	30	.14	.66
p value		n.s.	n.s.

normal subjects in the proporation of time they spent working on the two puzzles.

We would like to emphasize that we would not be loath to disproving a parallel, but we believe that this experiment was not a satisfactory analog of the typical test of passive-avoidance learning in animals. A conceptually equivalent test in humans would require a single response which leads to an attractive, positive consequence and at the same time, a painful, negative one. This and this alone would be the only available response alternative. Subjects must be impelled to approach a positive reinforcer and, at the same time, to avoid the noxious stimulus which always comes with the reinforcement. The subjects in this experimental attempt to test passive avoidance could work for the $.25 and no shock and in this way avoid altogether the approach-avoidance conflict of the $5.00 and shock puzzle—an option which in fact most subjects chose.

To summarize, for pain sensitivity and passive-avoidance behavior we have found no evidence supporting a parallel between obese humans and lesioned animals. We tend to believe that our experimental tests, particularly of passive avoidance, were inadequate. The reader may well feel, however, that this is self-indulgence and that there is no parallel for either pain sensitivity or passive avoidance.

PART II
ANIMAL–HUMAN BEHAVIORAL PARALLELS AND STIMULUS SENSITIVITY

Having presented this array of studies designed to test the parallels between the obese human and the VMH-lesioned animal, let us review the status of this comparison. From the previously existing research literature the following facts appear to hold for both the VMH-lesioned animal and the obese human:

1. The obese eat more good tasting food than do normals.
2. The obese eat less bad tasting food than do normals.
3. The obese eat on the average slightly, not hugely, more than normals do.
4. The obese eat fewer meals per day.
5. The obese eat more per meal than do normals.
6. The obese eat more rapidly.
7. The obese are less active than their normal counterparts.

From the experiments designed to test for humans what is already known about rats we know the following additional facts:

8. The obese are more emotional than are normals.
9. The obese do better at active avoidance.
10. The obese do not regulate intake in accordance with the caloric density of a solid preload or diet. Normals do regulate.
11. Both the obese and normals do regulate intake in accordance with the caloric density of a liquid preload or diet.
12. When obtaining food requires no particular effort, the obese eat more than normals do.
13. When it requires work to get at food and the food cue is remote, the obese eat less than normals do.

Of the variety of relationships studied, in only two cases are there indications that a parallel may not hold. Whereas studies on lesioned animals have indicated that the obese animal is more sensitive to electric shock, our own efforts to test this relationship on humans have yielded equivocal results. In one study, the obese did prove more sensitive to shock than normal controls; in a second study, there were flatly no indications of a difference between the two groups. Similarly, animal studies have indicated that VMH animals are worse at passive avoidance while our own efforts have, so far, failed to demonstrate such a relationship for humans.

All told, then, we have attempted to compare the behavior of the VMH-lesioned animal and obese human on 15 different facts. Though some of these facts are "obvious" in the sense of almost defining obesity—e.g., that the obese eat larger meals and more per day—most of the parallels are neither obvious nor trivial. In 13 of the 15 comparisons, the obese human and the lesioned animal behave similarly. All in all, this is, we believe, a formidable array of evidence indicating that there is *something* systematic and meaningful to be learned from this exercise in science by analogy.

On the most primitive level, of course, these striking parallels do suggest that there is something awry with the hypothalamus of the obese human. For reasons that have been spelled out elsewhere (Schachter, 1971b), we have deliberately avoided any frontal attack on the numerous physiological questions that are raised by these parallel sets of data but have chosen instead to attempt to create a coherent conceptual scheme that would allow us to integrate the diverse set of facts that seem, by now, to be well established for both lesioned animals and obese humans. Our efforts in this direction have been guided by the modifications that the data, both for rats and humans, have imposed on our original conceptions (Schachter, 1968, 1971a) about human obesity.

Our early studies (Schachter et al., 1968; Schachter & Gross, 1968; Nisbett, 1968b; Goldman et al., 1968) had indicated that the eating behavior of the obese is under external, rather than internal, control. When an external, food-relevant cue, presumably *any* food cue, is present, the obese are more likely to eat and to eat a great deal than are normals. When such a cue is absent, the obese are less likely to try to eat or to complain about hunger. In short, the obese are stimulus bound. That this presumed, generalized external sensitivity to food cues was an oversimplification became clear as we learned the following facts about the human obese:

1. The obese report eating less often per day than do normals. Since the world is full of food cues, a hypothesis of differential external sensitivity would virtually require that the obese eat more often.

2. In Nisbett's (1968a) 1- and 3-sandwich experiment, when they have only one sandwich in front of them, the obese eat significantly less than do normals despite the fact that they have been invited to help themselves to more sandwiches from the refrigerator across the room—clearly a food-relevant external cue.

3. In studies of fasting behavior (Brown & Pulsifer, 1965; Duncan, Jinson, Fraser, & Christori, 1962; Goldman et al., 1968) the obese apparently have an easier time fasting than do normals, despite the fact that in some of these

studies they do their fasting in the real world, again, a world full of food cues, each of which by hypothesis should be more likely to trigger an impulse to eat in an obese than in a normal person.

Short of invoking a variety of psychodynamic mechanisms and special considerations to cope with each of these divergent facts, it seems possible to reconcile our original view of matters with these facts only by invoking a new dimension—the prominence of the food stimulus—and by assuming that a food cue must be potent and compelling in order to trigger an eating impulse in an obese subject—the difference between a hot dog stand two blocks away and a hot dog under your nose, savory with mustard and steaming sauerkraut. In essence, this is an interactional hypothesis which assumes that the relationship of cue prominence to the probability of triggering an eating response is stronger for the obese than for normals. When the cue is immediate and compelling, the obese should be more likely to eat than normal subjects; when the cue is remote, the obese should be less likely to eat than normals.

Such a formulation does allow us to cope with virtually all of the known facts about the eating behavior of the human obese. This, however, is small triumph, for the hypothesis is an *ad hoc* construction designed to do just this. The following portion of the monograph is divided into two sections. The first section presents a series of experiments, designed directly to test the hypothesis and some of its implications for eating behavior. In the first experiment, Lee Ross (Chapter 7) presents fairly conclusive evidence that the basic hypothesis is correct. Ross manipulates cue prominence in two ways: first, by manipulating the sheer physical prominence or inconspicuousness of the food cue; second, by manipulating the extent to which the subject's attention is focused on the food cue. When the subject's attention is focused on a prominent food cue, obese subjects eat roughly twice as much as normal subjects. When the subject's attention is diverted from an inconspicuous food cue, the obese eat half as much as normal subjects.

Johnson's experiment (Chapter 8), which was done independently of our own research program, examines the impact of food-cue prominence on the subject's willingness to work for food. As would be expected from the interaction hypothesis, when food cues are prominent, obese subjects work considerably harder for food than normals do; when the food cues are remote, they work less hard than normals do.

Finally, the Schachter, Friedman, and Handler study (Chapter 9) is a field test of a corollary of the basic hypothesis. If it is correct that a prominent food cue is more likely to evoke an eating response from the obese, it could reasonably be expected that when faced with a plate of food, the obese would be unwilling to brook any obstruction or delay in getting at the food. It should be expected, then, that they will choose the easiest, quickest, most efficient means of eating. Observations in Oriental restaurants indicate that this is the case, since obese Occidental eaters are much less likely to use chopsticks and more likely to eat with their hands than are normals.

The final two chapters in this section are concerned with theoretical and analytic matters. In Chapter 10 we review the strengths and weaknesses of the cue prominence-reactivity hypothesis as it applies not only to the human data from which it evolved but to the body of experimental literature on the lesioned animal. Chapter 11 is a digression from the main theme of this volume and is an attempt to consider the question of causality as it applies to the repeated demonstration of an association between obesity and externality. By comparing dynamic to static lesioned animals we attempt to evaluate the effects of the lesion independent of obesity. By comparing formerly fat subjects to those who are presently obese we attempt to isolate the impact of current obesity.

7
EFFECTS OF MANIPULATING SALIENCE OF FOOD UPON CONSUMPTION BY OBESE AND NORMAL EATERS[1]

Lee Ross

The first experimental indications that the obese were more sensitive to environmental food-relevant cues and less sensitive to internal cues than normal subjects led to the hypothesis that the eating behavior of the obese is under external control—that is, an external, food-related cue is presumed more likely to trigger an eating or food-acquiring response in an obese than in a normal subject. The hypothesis implies that under any circumstances, if a food cue is present—no matter how remote or obscure the cue—the obese will be more likely to eat. Subsequent research quickly made it clear that in this form the hypothesis was simply wrong for the following reasons:

1. It should be expected from the notion of external control that under normal circumstances the obese will eat more frequently than normals; yet, in an eating-diary study, obese students report eating somewhat less frequently than do normal students (Ross et al., 1971).
2. It should be expected that the obese will have a harder time fasting than normals, yet this proves not to be the case (Goldman et al., 1968).
3. It should be expected that *any* food cue will be more likely to stimulate the obese to eat, yet as Nisbett's (1968a) one- and three-sandwich experiment demonstrated, the obese appear to eat less than normals when the food cue is remote or relatively inaccessible.

Obviously, the hypothesis of external control requires modification if we are to reconcile these contradictory facts with those which led originally to

[1]The author gratefully acknowledges the assistance of Dr. Stanley Schachter whose many helpful suggestions and criticisms contributed to this paper and to the completion at Columbia University of the doctoral dissertation upon which it is based.

the formulation of this hypothesis (Schachter, 1968). Since the circumstances under which the original, external-control hypothesis holds appear, in most cases, to involve particularly potent or salient cues—literally food under the subject's nose—we would suggest that the variable of cue salience may be crucial in delineating the conditions that distinguish obese from normal subjects.

The present experiment was designed to test the relationship of obesity to cue prominence or salience. If our speculations are correct, the obese should be expected to outeat normals when the food cue is particularly salient and not to do so when such a cue is remote. In this study, cue salience is manipulated in two ways: first, by varying an external property of food, that is, its visibility; second, by manipulating a subject's thought about, or attention to, food.

METHOD

Overview of Experimental Design and Procedure

Within the context of a study purporting to investigate the physiological correlates of thinking, obese and normal subjects were seated with ready access to a large container of cashew nuts. The salience of these nuts for subjects was manipulated by two different techniques. Cue salience was manipulated simply by varying the level of illumination in the room and hence the visibility and prominence of the nuts. A cognitive manipulation of salience was achieved through taped instructions which specified what subjects were to think about.

Subjects

There were 120 subjects in the study, all male Columbia undergraduates. Half of the subjects were obese, half were normal in weight. A subject was considered obese if he was 15% overweight according to the norms published by the Metropolitan Life Insurance Company. Obese subjects ranged from 15.3% to 56.7% overweight (mean +27.1%), and normal subjects ranged from 7.7% underweight to 10.0% overweight (mean +0.8%). A total of seven subjects were excluded because their weights placed them between the obese and normal cutoff points.

Experimental Procedures

Subjects were run in midafternoon (at 2:30, 3:30, or 4:30) when they presumably had finished lunch but had not yet begun to yearn for dinner.

Upon their arrival subjects were introduced to a "study of the physiological correlates of various types of thinking or cognitive activity." This deception allowed us to manipulate the subject's thinking and the conditions of his exposure to food without alerting him to our concern with eating behavior. It stressed that "amplitude and frequency of heartbeat" were chosen for the present study because of the "stability" and "robustness" of these measures, which allowed them to be tested "against a background of normal activity and movement."

The experimenter then demonstrated the recording apparatus, which consisted of a "miniature telemetric transmitter" and a "receiving console" located outside the experimental room, which the subject believed would record and automatically tally his cardiac response. When the subject was comfortably seated with the telemetric capsule in his shirt pocket and lead electrodes attached to his chest, the experimenter concluded his introductory remarks with a slightly expanded version of the following:

> As I explained before, we are going to ask you to think about various things so that we can measure the effects of different types of thinking or cognitive activity upon the amplitude and frequency of your heart beat. We use these objects you see on the table in front of you in this task.

The experimenter then touched and mentioned in turn each of the articles within the subject's easy reach. "There is the car, these chessmen, the candlesticks, the marbles, and these nuts." As the experimenter mentioned the nuts he removed the lid from the nut container to reveal 800 grams of fresh, salted, large cashews. The nuts were opposite the subject's right hand, the marbles opposite his left, and the car, chessmen, and candlesticks were directly in front of him. The introduction continued:

> The instructions for the thinking task will be delivered by the tape recorder here. It may tell you to think about one of these objects, it may tell you to think about anything you wish, or it may give you a thinking task that doesn't involve these objects at all. It depends upon which condition you are in, as different subjects will have different thinking tasks. Of course, you will be participating in two or three different thinking situations before we are through.
>
> Okay, we are just about ready to start now. I want you to be comfortable and relaxed. Just remain seated facing the table and don't smoke. Anything else is okay; there are no other restrictions since we want a technique that can be used in the real world where people can actively interact with the environment. You can fool around with the car, play with the marbles, eat the nuts, handle the chessmen, or the candlesticks, or do anything else you want. Okay, just relax and pay attention to the tape recorder. Remember you can eat, play, or do anything else you want as long as you remain seated and pay attention to the taped instructions.

It should be emphasized that the experimenter's descriptions and instructions were designed to make eating a perfectly acceptable behavior for the participants without making it an experimental requisite.

The Manipulation of Cue Salience

As the experimenter turned to leave he turned off the overhead lights in the room and turned on a small lamp, explaining that the reduced illumination would help the subject relax and produce a more stable heartrate record.

Strong Cues Condition: For half of the subjects the lamp was unshaded and contained a 40 watt bulb, which provided normal illumination. The visual cues provided by the large tin of cashews within the subject's reach were thus highly salient.

Weak Cues Condition: For the remaining subjects the lamp was shaded and contained a 7½-watt red bulb, which drastically reduced the illumination in the room. While the tin of nuts and the other objects could still be readily detected and reached for, the potency or salience of the visual cues they provided was greatly reduced in comparison with the Strong Cues condition.

The Manipulation of Cognitive Salience

As the experimenter left he turned on a tape recorder which first reiterated some of the experimenter's explanations and instructions and then delivered instructions to the subject in order to manipulate his thinking during the experimental session.

High Cognitive Salience Condition: In this condition, obese and normal subjects received the following instructions:

> For the next few minutes as we adjust the recording apparatus we would like you to think about cashew nuts like those on the table in front of you right now. Relax, and think for a while about the cashews . . . about seeing them, think about eating cashews . . . think about this for a while until you hear the tape again.

This initial communication lasted 16 seconds and was followed immediately by 44 seconds of silence until the second communication began. In all, eight communications were presented, each lasting 15-20 seconds followed by 40-45 seconds of silence. Each communication focused on one or more of the properties of fine cashews, stressing the taste, texture, and aroma of the nuts and urging the subject to keep concentrating on these nuts. The communications were designed to keep the subject thinking about cashews, a cognitive manipulation designed to make the nuts highly salient. We term this condition the High Cognitive Salience condition (*High Cog*).

The second of the eight communications, typical of those delivered in this experimental condition follows:

> Now, think about the taste of cashews. Sit back and think of the light salty taste, the rich toasted quality, and the taste of really fine cashews. What other adjective describes that rich yet elusive taste? Can you think of any? Try, as you continue to think about the cashew for the next little while.

Low Cognitive Salience Condition: Subjects in this condition received eight communications practically identical in length, duration, and structure to the eight communications described above. The communications were designed to focus attention on something other than cashews, so that the nuts would not be salient for these subjects. Thus for subjects in this Low Cognitive Salience condition (*Low Cog*), the communications dealt with the sensual properties of marbles. The following are the first and the last of these eight communications:

> For the next few minutes as we adjust the recording apparatus we would like you to think about marbles like those on the table in front of you right now. Relax and think for a while about the glass marbles, about seeing them . . . handling them, playing with them. Think about this for a while until you hear the tape again.
>
> All right, the apparatus should be tuned in now. Just a little while longer, keep thinking about the bright, multi-colored marbles, creamy white, lustrous, rich reds, greens, blues, shining clear hard glass, the satisfying clinking sound they make in contact. A little while longer please, relax, and keep thinking about the marbles.

Free Thinking Condition: The remaining subjects also received eight communications of similar length, duration, and structure to those already described for the *High Cog* and *Low Cog* conditions. For these subjects, however, the communications were designed to allow them total freedom in determining what they wish to think about. We have labelled this condition the *Free Think* condition, since participating subjects were free to concentrate their thinking on the cashews to whatever degree they wished.

To achieve maximum comparability between this condition and the *High Cog* and *Low Cog* conditions the subject was again presented with eight communications. The two communications presented below are typical of the *Free Think* communications:

> In the next few minutes as we adjust the recording apparatus you may choose any object or event to think about you wish. You may change the subject of your thoughts anytime you want to but please try to keep thinking about only one thing at a time . . . So just relax and keep thinking about one thing at a time until you hear the tape again.
>
> Are you having difficulty thinking as we have asked you, or do you find it easy to relax and savor the details of your thoughts, the details of sight, sound, smell, or whatever is appropriate to your thoughts? Think about how the object or event affects your senses. Try to keep thinking in this way.

Measurement of Cashew Consumption

When the eight communications were completed the experimenter immediately escorted the subject from the room so that an assistant could weigh the container and record the subject's consumption of cashews.

Concluding Procedures: There followed two additional sessions during which the effects of further manipulations upon eating and other behavior were studied. These sessions are described in detail (Ross, 1969) in an earlier and more comprehensive research report. Where relevant to present concerns, the results of these additional sessions clearly support the conclusions offered in this paper. At the conclusion of the study the height and weight of participants were measured and their ages recorded. Final debriefing for all subjects occurred at the end of the academic semester.

RESULTS

Table 1 indicates the mean consumption in grams[2] for subjects who received each of the cue and cognitive manipulations of salience.

Effects of Cue Salience Manipulations

It is immediately clear that the obese subjects were considerably affected by the cue manipulations. Overall, Strong Cue obese subjects consumed twice as much as Weak Cue obese subjects. Subjects of normal weight provide a sharp contrast as they consumed fewer grams when exposed to Strong than when exposed to Weak Cues.

We may express the same interaction by noting that obese subjects ate considerably more than normal subjects when cues were manipulated to make the cashews highly salient but ate less than normal subjects when low salience conditions prevailed. When obese subjects are considered separately in an analysis of variance, the main effect of the cue manipulation is clearly significant ($F = 7.85$; $p < .01$). When the eating of normal subjects is similarly analyzed this effect is negligible ($F < 1.00$). When the two weight groups are considered within a single analysis the interaction between weight and the cue manipulation of salience is statistically significant at the .02 level ($F = 6.80$).

With respect to cue manipulations of salience the main experimental hypothesis is clearly supported. The heightened sensitivity or responsiveness of obese subjects seems restricted to conditions where prominent or strong cues make food highly salient for these subjects. Normal eaters, in contrast to obese eaters, responded equally to strong cues and weak ones.

[2]The cashew nuts used in this study weighed between 1 gm and 1.5 gms each.

TABLE 1

Grams of Cashews Consumed under High and Low Salience Conditions

	Obese subjects							
	Cognitive salience manipulation							
Cue salience manipulation	*High Cog*		*Free Think*		*Low Cog*		Combined	
Strong cue condition	47.2	(10)[a]	24.2	(7)	39.1	(9)	36.9	(26)
Weak cue condition	31.5	(10)	11.3	(7)	13.7	(5)	18.8	(22)
Combined	39.4	(20)	17.8	(14)	26.4	(14)	27.9	(48)
	Normal Subjects							
	Cognitive salience manipulation							
Cue salience manipulation	*High Cog*		*Free Think*		*Low Cog*		Combined	
Strong cue condition	24.1	(10)	17.5	(6)	18.7	(7)	20.1	(23)
Weak cue condition	20.9	(7)	28.7	(7)	25.1	(7)	24.9	(21)
Combined	22.5	(17)	23.1	(13)	21.9	(14)	22.5	(44)

[a]Bracketed figure indicates the number of *S*s in each condition who ate at least one cashew. Means, however, are always based on the consumption of all ten subjects in each experimental condition.

Effects of the Cognitive Salience Manipulations

The cognitive manipulation, like the cue manipulation, seems to have exerted considerable influence on obese subjects. When free to think about anything they wish (Free Think condition) obese subjects consumed an average of less than 18 grams of cashews during the eight-minute period. When directed to think about cashews in the *High Cog* condition, overweight subjects consumed a mean of more than 39 grams of nuts within a similar eight-minute session. Interestingly, the *Low Cog* condition produced an intermediate amount of eating; overall, these subjects consumed an average of 26.4 grams of nuts as they sat thinking about marbles. It should be noted that obese *Low Cog* consumption is particularly great under Strong Cue conditions. These obese subjects, in fact, eat almost as much as obese subjects who were exposed to *High Cog* instructions combined with Strong Cues—an unexpected finding which suggests that for some reason the obese are particularly vulnerable to prominent cues when their attention is distracted from such cues.

Once again, an examination of eating by normal subjects in the relevant experimental conditions provides a sharp contrast to the pattern of experimental effects demonstrated by obese eaters. Quite simply, normal subjects seem to have been impervious to the cognitive manipulation as they consumed an average of slightly more than 20 grams regardless of their cognitive salience condition.

Statistical analysis reveal that when obese subjects are considered separately, the main effect of the cognitive manipulation is significant ($F = 3.82$, $p < .05$). More specifically, the *High Cog* condition produced significantly greater cashew consumption than the *Free Think* condition ($F = 6.89$, $p < .02$) while other paired comparisons yielded only marginal significance. When normal subjects are considered separately, none of the tiny differences in mean consumption among the three cognitive salience conditions approached statistical significance.

Overall, the interaction between weight and cognitive salience did not reach accepted levels for statistical significance ($F = 2.16$, $p < .13$). A significant interaction between weight and cognitive salience occurs when *High Cog* and *Free Think* consumption is contrasted for obese and normal subjects ($F = 4.09, p < .05$).

DISCUSSION

The present experiment supports the hypothesis that obese subjects are more responsive than normal subjects to food-related cues only when such cues are highly salient or prominent for the subject. Thus obese subjects *are* "particularly sensitive" or oversensitive in terms of the magnitude of their response to highly salient food; they *are not* particularly sensitive in terms of the level of stimulation required to initiate appreciable eating. In fact, in condition where subjects faced weak cues and were not specifically induced to think about food, obese subjects consumed considerably less than normal subjects.

This pattern of results helps us understand why obese individuals eat larger but fewer meals than normal individuals (Ross et al., 1971) and are inveterate plate-cleaners (Nisbett, 1968a). Immediately available food triggers excessive consumption by the obese individual; most food-related cues present in advertisement, conversation, and the sights and sounds of everyday life he resists quite successfully. Similarly we can understand why obese subjects, in the Goldman et al. (1968) study of religious fasting, experience very little discomfort when they remain in the synagogue shielded from all palpable or compelling food-related cues.

With respect to the cognitive manipulations the data are less clear-cut in their implications. Overall, obese subjects ate more under *High Cog* than under *Free Think* or *Low Cog* conditions while normal subjects were apparently uninfluenced by the salience manipulation. But there is one anomaly in the data: Obese subjects exposed to strong cues ate more under *Low Cog* than under *Free Think* conditions. While this second order interaction among weight, cue strength and cognitive salience is not statistically significant, it does suggest that the effects of the two salience manipulations were not simply additive. The possibility that distraction may increase the obese eater's susceptibility to strong or salient food-related cues merits further investigation, for it has obvious implications for the problem of weight control.

8

THE EFFECTS OF CUE PROMINENCE AND OBESITY ON EFFORT TO OBTAIN FOOD[1]

William G. Johnson

When food is freely available and no effort is required to obtain it, ventromedial-lesioned rats eat more than controls. When work is required to get food, the lesioned animals exert less effort than controls and consequently eat less. On the basis of such experimental findings Miller et al. (1950) caution against "drawing inferences about 'drive' from consummatory behavior . . . [p. 259]." It would appear that consummatory and instrumental food-related behavior do not necessarily vary in the same fashion. This conclusion is based on experiments that manipulate the difficulty of obtaining food by devices such as varying the number of bar presses required to get a pellet or measuring the speed of running down an alley to obtain food.

It is the case that virtually all demonstrations of the unwillingness of the lesioned animal to work for food have confounded effort and cue prominence. Food at the end of an alleyway is obviously a more remote cue than food in the animal's cage. A high FR schedule not only increases the effort of the task but also decreases the salience of the food cues. Given the potent effects of cue prominence demonstrated in experiments such as those of Ross (Chapter 7), Rodin (Chapter 13), and Pliner (Chapter 14), it is conceivable that these effects of effort may be explainable in terms of cue prominence.

If this is correct it should be expected that the obese will be less willing than are normals to work for food when food cues are remote, and more

[1]This study is a portion of the research submitted in partial fulfillment of the requirements for the Ph.D. degree at the Catholic University of America. The author wishes to thank Richard A. Wunderlich, Clement Gresock, and Robert Dowling for their assistance as members of the dissertation committee. The research was supported in part by Grant # RD-224-M-69 from the Social and Rehabilitation Service of the Department of Health, Education and Welfare.

willing to work when the food cues are prominent. It is the purpose of this experiment to test these expectations on human subjects.

METHOD

Procedure and Design

The experiment was conducted under the guise of a study investigating the relationship between taste and effort. At the time subjects were recruited, they were asked to abstain from food for at least four hours prior to their appointment and for at least two hours after the experimental session. Since a free lunch was the incentive for participation in the experiment, subjects' preference for either roast beef, turkey, or baked ham was obtained.[2] After the subjects indicated their choice, they were asked to state how well they liked that particular type of sandwich. All subjects verified that the chosen sandwich was among their most preferred over all others.

Normal and obese subjects were randomly assigned to one of four experimental conditions: (*a*) Prior Taste-Food Visible (PT-FV), (*b*) No Prior Taste-Food Visible (NPT-FV), (*c*) Prior Taste-No Food Visible (PT-NFV), and (*d*) No Prior Taste-No Food Visible (NPT-NFV). All subjects came to the laboratory between 11:00 A.M. and 2:00 P.M. and, upon arriving, were asked about their prior eating. All subjects reported a food deprivation period of at least four hours.

The experimental instructions were presented on tape and began:

> A subject of considerable importance in psychology today is taste, and as you know, this experiment involves the relationship between taste and effort. Some research has indicated that the harder one works for a reward, the more valuable that reward is for him. For instance, a person who receives a reward for a difficult task will tend to place greater value on that reward than if he received it for working at a lesser task.
>
> This experiment will attempt to determine if this relationship holds for taste. That is, will working at a moderately difficult task for which food is received, dispose one to rate the food as tasting better than if the food were received for less effort.

After an explanation of why they had been asked not to eat prior to coming to the laboratory, the following "effort" instructions were given:

> Now look at the apparatus. See the ring with the attached wire, the light directly above it, and the opening in the console which is enclosed by the plastic case. The case now contains a sample of

[2]These sandwiches were selected on the basis of a survey of the five most preferred sandwiches in a sample of 25 individuals. In addition to the above, corned beef and a ham-and-cheese combination were included as choices for the first 30 subjects who were tested. These latter choices were selected infrequently and subsequently dropped.

> the type of sandwich which you indicated as desirable. Since you have not eaten for some time and have agreed not to eat for at least two hours after the experiment, we want you to earn your meal now. The task consists of pulling the ring with your index finger for 12 minutes. As you pull the ring, the light will go on—you must pull it back far enough to turn the light on in order to get credit. You will receive portions of your desirable sandwich through the opening as you pull the the ring. The sandwich portions are very delicious, delicatessen type. After your 12 minutes of working time you will taste and rate the portions and have the opportunity to eat only what you have earned.

Following the taped instructions, questions were answered as the experimenter adjusted the apparatus. The preferred hand was placed in a restraint, and the index finger was inserted through the ring.

Task Difficulty: The task was defined as a moderately difficult one in terms of both the effort involved in pulling the ring and the relationship of responses to food portions. In order to assure moderate effort for all subjects, the maximum weight at which each subject could close the circuit (turn on the light and activate a response counter) was determined to the nearest half-pound. The working weight was then one-half of the maximum. For example, if a subject could close the circuit at 15 lbs. but not at 15.5 lbs., his working weight was set at 7.5 lbs. The mean working weights were 7 lbs. for normals and 7.5 lbs. for the obese.

A VR/50 schedule relating responses to food portions was used.

Prior Taste Manipulation: Half the subjects consumed a one-quarter portion (approximately 20 grams) of desirable food immediately before commencing the task. All subjects reported that the portions tasted good-to-excellent. The other subjects consumed portions of plain white bread (approximately 15 grams).

Food Visible Manipulation: For half the subjects, one sandwich of desirable food attractively presented and covered in a *transparent wrap* (Glad Wrap) was placed to their left. Smaller portions of desirable food (one-quarter sandwiches, approximately 20 grams) similarly wrapped were delivered as the VR requirements were met; however, they could not be eaten until after the 12-minute working period.

For half the subjects, the portions were wrapped in *white nontransparent* shelf paper. For convenience, pieces of foam identical in size and shape were substituted for the desirable food. Subjects were never exposed to these fake portions and were given every reason to believe that the fake portions were actually the preferred food.

Following the working period, the subjects were directed to another room where they were given their earned portions and a taste-rating scale. After finishing their meal, the actual height and weight of each subject was obtained.

TABLE 1

Mean Responses as a Function of Weight and Cue Prominence

Conditions	Subjects: Obese	Subjects: Normals
Double cue		
Prior taste/food visible	591.0	371.8
Single cue		
No prior taste/food visible	561.6	343.9
Prior taste/no food visible	469.3	445.5
Minimal cue		
No prior taste/no food visible	352.4	439.8
Combined	493.6	400.2

Subjects

Eighty subjects, recruited from graduate and undergraduate courses, participated in the study. The average age of the subjects was 32.8. The obese group contained 31 females and 9 males while there were 29 females and 11 males in the normal group. These subjects were proportionately distributed across the eight treatment conditions. Overweight subjects ranged from 14.3 to 102.6% overweight, with a mean of +36.5%. Normals ranged in weight from −14.0 to +7.0% overweight with a mean of −0.2%.

Apparatus

The apparatus was mounted on a black table. A plastic covered metal ring (1/8 in. in diameter) was located approximately two feet from the subject's end of the table and was attached to 12-gauge wire. The wire ran underneath a large black plywood panel which faced the subject and stood upright and perpendicular. The wire was strung over a pulley at the end of the table and down toward the floor where adjustable weights were secured. A microswitch and a signal light were connected to the wire. The circuit would close with approximately a 1-in. pull on the metal ring. Closing the circuit turned on the white signal light which was mounted directly above the wire and at eye level to the subject as he sat before the table.

To the subject's left was a transparent plastic case which was flush with the plywood panel. The case enclosed the sample sandwich and storage space for

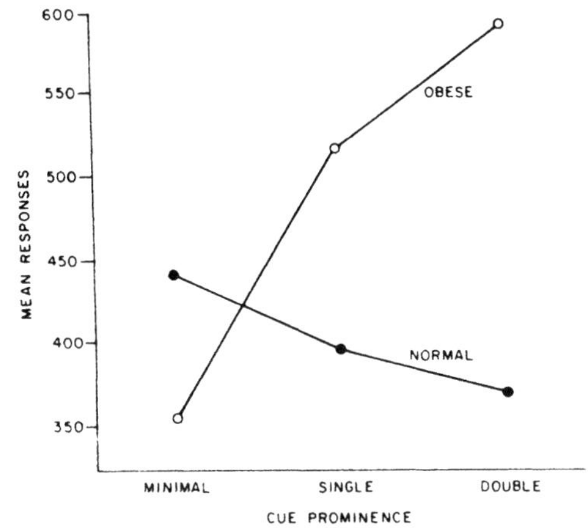

Fig. 1. The effect of weight and food cue prominence on food-directed performance.

the earned sandwich portions. The portions were delivered through a chute in the case from the rear side of the panel by the experimenter, according to the VR/50 schedule.

The number of times the subject flexed his index finger with the attached working weight was recorded on a response counter.

RESULTS

It was predicted that overweight subjects would have fewer responses than normals under conditions of low cue salience and more when food cues were highly prominent. It can be seen from the means presented in Table 1 that when desirable food is both highly visible and recently sampled by the subjects (double cue condition), the obese do in fact make a significantly greater number of responses than normals ($t = 2.87$, $df = 18$, $p < .02$).[3] Under minimal cue prominence, the obese make an average of 87.4 fewer responses than normals in the same condition. This difference is consistent with expectations; however, it is significant at only a peripheral level of confidence ($p < .20$).

Figure 1 illustrates the relationship between weight and cue prominence observed in the present study. The midpoint of the curve representing the single cue condition is an average of the PT and FV treatments for the obese and normal groups taken separately. As was predicted, there is a positive relationship between the mean number of responses and cue prominence for the obese subjects. The linear component of this curve is significant at the .005 level ($F = 8.96$, $df = 1/72$). Similarly, according to expectations the corresponding curve for normal subjects is stable over variations in cue prominence (linear $F < 1$, $df = 1/72$).

[3] All p values are based on two-tailed tests.

To determine the relative contributions of the prior-taste and food-visibility manipulations, a three-way analysis of variance for these factors and weight was carried out. Results indicate a significance weight × food-visible interaction ($F = 9.87$, $df = 1/72$, $p < .005$) and a nonsignificant interaction of weight and prior taste ($F < 1$). It is clear that the differential performance of the obese in response to cue salience is due almost entirely to the food-visible treatment.

DISCUSSION

The present study investigated whether the food-directed performance of the human obese was differentially responsive to food cue prominence. It appears that, as in Ross's experiment (Chapter 7), cue salience has a marked effect on the obese, for they work much harder to obtain food when the cues are prominent than when they are remote. In striking contrast the performance of normal subjects is relatively unaffected by cue prominence.

The relatively low effort with minimal food cues, while not statistically significant, is nevertheless suggestive as it corresponds with the results of both the lesioned-rat experiments (Miller et al., 1950; Teitelbaum, 1957) and Schachter and Friedman's nut experiment (Chapter 2). As the food cues are augmented the results are in line with the prediction that the performance of the obese would be increased. When the food was directly visible, overweight subjects performed a relatively difficult task to obtain food. This finding supports Nisbett's (1968a) contention of the instigative potency of food visibility in the human obese. It would be interesting to see if this hyper-responsiveness to visible food cues is also typical of obese rats. If so, the low level of performance shown in the Miller et al. (1950) and Teitelbaum (1957) studies might not have beeen the case if the food were visible.

The influence of food visibility is important in view of the finding that prior taste of the food seemed to have very little effect on the performance of the obese. Certainly it seems reasonable to expect that nothing could be more salient or potent than food in one's mouth. The lack of a significant relationship between weight and prior taste was unexpected and appears difficult to reconcile with previous work. However, recognizing the power of visible food in controlling the eating behavior of the obese, it is not as surprising that a prior taste has less effect than visible food. In fact, the PT condition and its effect on effort in this case is markedly similar to Nisbett's (1968a) one-sandwich condition and its effect on food consumption. When this single sandwich had been eaten and thus was no longer directly visible, the obese ate less additional food than did normals. In the prior taste condition in the present study, consumption of the sandwich portion similarly removed the effective cue for enhanced effort. In light of this consideration, it is likely that the difference between the prior-taste and food-visible conditions is not solely one of taste vs. vision, but also between a cue which is immediately available and one which is more remote.

This analysis of the results suggests, then, that the food-visible condition successfully unconfounded effort and cue prominence while the prior taste condition did not. A simple test of whether this is true would be to reverse the immediacy of the food cues in each condition as they relate to performance. If the FV condition were changed so that the subjects saw food portions prior to but not while working and in the PT condition subjects tasted food while working, not before, we would expect the empirical relationships to reverse.

Finally, the present experiment allows us to examine the effects of cue salience in a context where, as in a person's natural environment, a series of performances are required in order to eat. Typically, food must be obtained and/or prepared to some degree before eating can begin. The present study demonstrates that, for the obese, the prominence of the food cues affects not only how much food is eaten, but how much effort they are willing to make in order to obtain food.

9
WHO EATS WITH CHOPSTICKS?

Stanley Schachter, Lucy N. Friedman,
and Joel Handler

The evidence is compelling that prominent food cues have particularly potent effects on the eating behavior of obese subjects. In Chapter 7 Ross has demonstrated that the obese will eat far more from a brightly illuminated bowl of nuts than from a dimly illuminated bowl, whereas the degree of illumination has little effect on the amounts eaten by normal subjects. Johnson, in Chapter 8, has shown that in the presence of multiple food cues, the obese will work far harder for food than in the absence of such cues and, again, normals are relatively unaffected by the presence or absence of food cues. Nisbett (unpublished data cited in Chapter 1) has found that when presented with a bowl of ice cream, obese subjects eat more rapidly than do normals.

Taken together, this variety of facts would certainly suggest that the sight of potent food cues serves to strongly motivate obese subjects to eat. If this is correct, it seems reasonable to assume that, in the presence of food, the obese will be loath to tolerate delay or obstruction in getting the food down. If the food is good, they should choose the quickest, easiest, most efficient means of eating. Though in public one rarely has the option of choosing among alternative modes of eating, there is one institution in which there is always a choice—the Chinese restaurant where one may eat with chopsticks, or with fork and spoon. Most Western eaters, of course, are completely adept with silverware and clumsy and inept with chopsticks. It seems at least reasonable then to guess that when faced with a plate of Chinese food, the obese Occidental will be less likely to eat with chopsticks than will his normal counterpart. To test this casual derivation, the investigators visited Oriental restaurants and simply observed the table manners of the patrons.

Procedure

As standard procedure, two observers went into a Chinese or Japanese restaurant approximately 15 minutes before the usual lunch or dinner time. They chose a centrally located table, ordered a meal, and unobtrusively drew floor plans of the restaurant. As each customer entered the restaurant, each observer independently categorized him as normal, obese, or chubby. They also noted the patron's sex and race and estimated his age, height, and weight. All such observations, of course, were made before the customer ate.

Once the meal was served, the observers simply noted whether the customer ate with chopsticks or silverware. In 11 of the 16 restaurants visited, the number of customers was small enough so that the observers could continually watch the eating habits of all patrons. In those restaurants in which all subjects could not be constantly watched, the observers made at least three checks on each eater to determine if he switched to or from chopsticks during the meal. In addition to this basic observation, in a sample of the restaurants visited, the observers systematically noted whether or not a customer, at some point during the meal, ate with his hands.

The Restaurants: During the course of the study sixteen visits were paid to a total of fourteen different restaurants—twelve of them Chinese and two Japanese. The restaurants were all on the upper West Side of Manhattan within roughly a mile of Columbia. For 12 of the 16 visits, the tables were set with silverware and it was necessary to ask the waiter for chopsticks. On two occasions—both buffets—chopsticks were placed alongside the silverware. In two restaurants, chopsticks were set on the table and one had to ask for silverware.

During the 16 visits, a total of 493 eaters were observed. Of these, 454 were Westerners and 39 were Orientals.

Judging degree of obesity: To train themselves in judging body size before the study began, both observers independently categorized and then estimated the weight and height of 35 undergraduates, graduate students, and faculty members at Columbia all of whom were subsequently weighed and measured. Calculating weight deviations from the Metropolitan Life Insurance Co. (1959) tables of average weight and height, the two observers differed from one another in calculated percent-weight deviation by an average of 5.5% and differed from the truth by an average of 6.2%. Following the rules of the classification system used for the subjects in the restaurants proper, only one of the 35 practice subjects would have been assigned to the wrong category.

In the restaurants, the observers first categorized a subject as obese, normal, or chubby and then estimated height and weight. Subjects were assigned to weight-deviation category by the following set of rules:

1. If both observers classified a subject as obese, normal, or chubby (93.2% of all cases) he was counted as such regardless of height and weight

estimates. When the two observers disagreed on classification, the subject was classified according to the following rules:

2. If one observer classified a subject as obese, and the other called him normal, the subject was considered chubby (2.0% of all cases).

3. If one observer called a subject chubby, and the other observer categorized him as either obese or normal, the disagreement was resolved by the use of the height and weight estimations (4.8% of all cases).

In estimating weight deviations, the criterion for obesity in this, as in most of our other studies, is 15% or more overweight. Anyone less than 10% overweight is considered normal, and people falling in the no-man's land between these cut-off points are classified as chubby.

Results

Considering each visit to a restaurant as an independent test of the hypothesis, we have a total of 16 replications. Of these 16 restaurants, 4 were no-test situations, since there were no obese customers during our visit. In 11 of the remaining 12 restaurants, the proportion of normally-sized, Occidental eaters using chopsticks was higher than that of obese eaters. In one restaurant, the hypothesis was not confirmed, since no Westerner, fat or normal, used chopsticks. By binomial test an 11 to 1 distribution is significant with $p < .01$.

The pooled data[1] for all 16 restaurants is presented in Table 1 where again it can be seen that normal eaters of Chinese food are far more likely to use

TABLE 1

The Relationship of Weight Deviation to Chopstick Use

Weight classification	Number of subjects who used	
	Silverware	Chopsticks
Normal	299	67
Chubby	36	0
Obese	41	2

χ^2 Normals vs. Obese = 4.19; $p < .05$.
χ^2 Normals vs. Chubby = 6.65; $p = .01$.

[1]These data do not include the 9 people (5 normal, 1 chubby, 3 obese) who tried both silverware and chopsticks. It was our impression that most of these people at some point during their meal experimentally tried out chopsticks.

chopsticks than are chubby or obese eaters. Contrapuntally, we note that among Orientals, for whom presumably silverware is the ordeal, there is a reverse tendency. Whereas 100% of fat and chubby Orientals ($N = 6$) used chopsticks, 91% of normal Chinese and Japanese ($N = 33$) did so.

Given these indications that the obese Westerner is less likely to use chopsticks, it could be expected that there will be other indications in his eating habits, that when faced with food, the obese eater will go about his job quickly, efficiently, and without fooling around with foreign implements. In four of the restaurants we systematically observed whether the customers used their hands while eating. We exclude from consideration the consumption of finger foods such as shrimp toast or barbecued spare ribs, which are designed and intended to be eaten with the hands, and include only instances of using the hands to get down such things as rice and tidbits of chicken, fish, mushrooms, etc. In these four restaurants, 42.1% of the obese (8 of 19) used their hands and only 2.8% of normals (4 of 142) did so—a difference significant by χ^2 test with $p < .001$—a finding which not only supports our general line of reasoning but disposes of the possible alternative explanation of the chopstick data that the obese are more sensitive about food and table manners and do not wish to appear clumsy or crude while eating.

The indications are consistent and strong, then, that when faced with a prominent food cue, the obese will choose the quickest and most efficient means of eating.

10
EXTERNAL SENSITIVITY AND THE VMH-LESIONED ANIMAL

The studies reported in the previous section have indicated that though the human obese are indeed more sensitive to external food cues than are normals, this effect is limited to potent and compelling food cues. A remote food cue seems less likely to evoke a food-acquiring response from an obese than from a normal subject. Further, there are indications that these relationships are not limited only to eating behavior but do generalize to other behavioral domains, such as emotionality. Figure 1 presents the postulated relationships in their most general form. For any cue, a prominent stimulus[1] is more likely to trigger an appropriate response from an obese than from a normal subject.

Though there are problems, on the whole, this formulation fits what we know remarkably well.[2] For human eating behavior, virtually every fact we

[1]By stimulus prominence we refer broadly to those stimulus or stimulus-field properties which compel or attract a subject's attention. We are deliberately sidestepping any attempt to define the dimension more precisely, for to do so at this point would, it seems to us, be premature and misleading. On the most primitive level one could, of course, define the concept in terms of the physical properties of the stimulus, so that as in Ross' (Chapter 7) experiment a brightly illuminated stimulus is considered more prominent than a dimly illuminated stimulus or a powerful smell more prominent than a subtle one. We assume it is obvious that any such definition would be grossly simplistic, for certainly such stimulus properties as novelty, unpredictability, the relationship to other elements in the field, etc. play at least as important a role in compelling attention as any purely physical stimulus property such as brightness or loudness.

[2]With one persistent and perplexing exception—as drawn, Fig. 1 suggests a stronger positive relationship between stimulus prominence and reactivity for the obese than for normal subjects, but for both groups the relationship is positive. We draw these curves in this fashion because it seems simple common sense to assume that for normals, too, cue prominence will have, at least, some effect. However, the data consistently belie common sense, for in study after study the relationship is slightly negative for normals. In Johnson's study, normal subjects work somewhat harder when there are no food cues than when there are multiple cues; in Ross's study, normal subjects eat somewhat less when the nuts are brightly illuminated than when they are dimly illuminated; in Nisbett's experiment (1968a), normals eat very slightly more when there is one sandwich before them than when there are three, and so on. At this point, we have no satisfying explanation for this perverse phenomenon. Even as obvious a suggestion as the possibly repellant nature of too many food cues won't do, for as in Rodin's study (Chapter 13), this negative relationship also seems to hold in experiments having nothing to do with food.

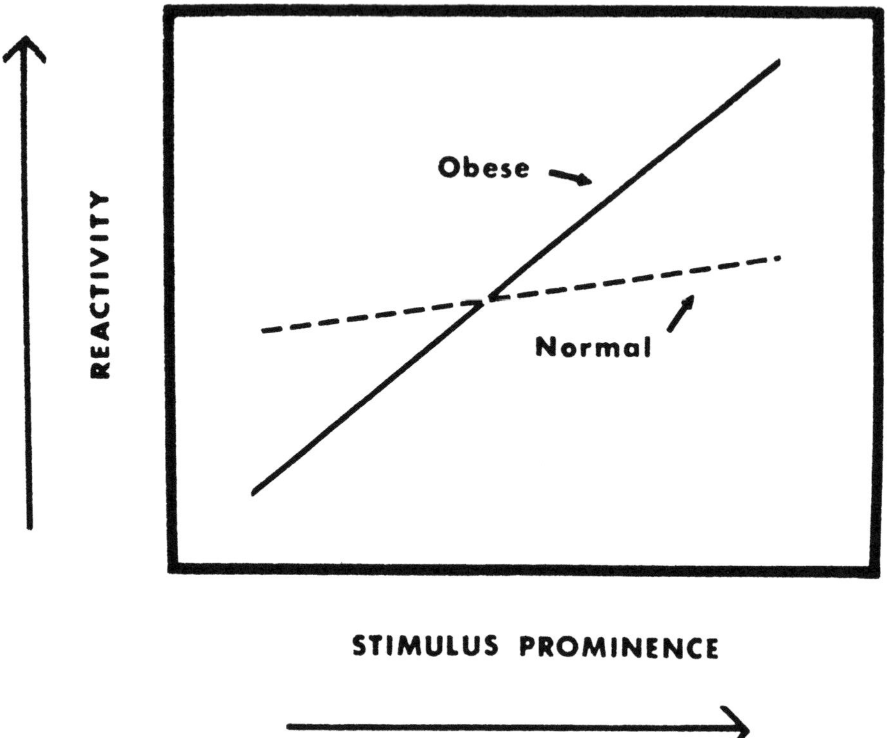

Fig. 1. Theoretical curves of relationship of reactivity to stimulus prominence.

have supports the assertion that the obese eat more than normals when the food cue is prominent and less when the cue is remote. Ross's (Chapter 7) experiment, of course, directly demonstrates the phenomenon. In Johnson's (Chapter 8) study of work and cue prominence, the obese do not work as hard as normals when there are no prominent food cues, but work much harder when the food cues are highly salient. In Nisbett's (1968a) 1- and 3-sandwich experiment, the obese subjects eat just so long as food cues are prominent—i.e., the sandwiches are directly in front of the subject—but when these immediate cues have been consumed, they stop eating. Thus, they eat more than normals in the 3-sandwich condition and less in the 1-sandwich condition. We also know from our fasting studies (Goldman et al., 1968) that the obese have an easy time fasting in the absence of food cues and a hard time in the presence of such cues, and so on.

About eating habits we know that the obese eat larger meals (what could be a more prominent cue than food on the plate?), but eat fewer meals (as they should if it requires a particularly potent food cue to trigger an eating response). Even the facts that the obese eat more rapidly and choose the easiest means of eating can be easily derived from this formulation.

About "non-eating" behavior we have less information (in the material presented so far) but the findings on emotionality, at least, seem to fit this formulation. In the Rodin experiment (Chapter 3) which required subjects to listen to either neutral or emotionally disturbing tapes, obese subjects described themselves (and behaved accordingly) as less emotional than normals when the tapes were neutral and much more emotional than normals when the tapes were disturbing. All in all, a surprising number of quite diverse facts about obese and normal humans can be subsumed under this relatively simple schema.

Let us return now to the VMH-lesioned animal. The striking set of parallels uncovered by our exercise in "science through analogy" does, it seems to us, legitimize what might otherwise seem a totally frivolous set of questions. Can the facts about the lesioned animal be explained in terms of these notions about stimulus sensitivity? Does this schema evolved from a comparison of obese and normal humans have anything to tell us about the ventromedial-lesioned rat? Before confronting these questions directly, let us consider current hypotheses about the lesioned animal. Theorizing about this preparation has tended to be modest and cautious and has been confined largely to speculation about the functions of the ventromedial hypothalamus. Essentially two suggestions have been made—one that the area is a satiety center, and the other that the area is an emotionality center. Anand and Brobeck (1951), Miller (1964), and Stellar (1954) have tentatively suggested that the ventromedial area is a satiety center—that in some fashion it monitors the signals indicating a sufficiency of food and inhibits the excitatory impulses initiated in the lateral hypothalamus. This inhibitory-satiety mechanism can account for the hyperphagia of the lesioned animals and, consequently, for their obesity. It can also account for most of the facts outlined earlier about the daily eating habits of these animals. It cannot by itself, however, account for the finickiness of these animals, nor can it account for the apparent unwillingness of these animals to work for food when they have been deprived. Finally, this hypothesis is simply irrelevant to the demonstrated inactivity and hyperemotionality of these animals. This irrelevance, however, isn't critical if one assumes, as does Stellar, that discrete neural centers, also located in the ventromedial area, control activity and emotionality. The satiety theory, then, can account for some, but by no means all, of the critical facts about eating, and it has nothing to say about activity or emotionality.

As a theoretically more ambitious alternative, Grossman (1966, 1967) has proposed that the ventromedial area be considered the emotionality center and that the facts about eating be derived from this assumption. By definition, Grossman's hypothesis accounts for the emotionality of these animals, and his own work on active avoidance learning certainly supports the emotionality hypothesis. It is, however, difficult to understand just why these emotional animals become fat. In essence, Grossman (1967) assumed that "ventromedial damage . . . . increase(s) an animal's emotional responsiveness to all stimuli. [p. 358]."[3] On the basis of this general statement he suggests that "the 'finickiness' of the ventromedial animal might then reflect a change in its affective response to taste [p. 358]." This could, of course, account for the fact that lesioned animals eat more very good-tasting and less very bad-tasting food than do normals. However, it is difficult to see just how this affective hypothesis can account for the basic fact about these animals—that for weeks on end the lesioned animals eat grossly more of ordinary, freely available lab chow.

Grossman (1967) attributes the fact that lesioned animals won't work for food to their "exaggerated response to handling, the test situation, the deprivation regimen, and the requirement of having to work for their daily bread [p. 358]," all of which is possible, but does seem far-fetched. At the very least, the response to handling and to the deprivation regimen should be just as exaggerated whether the reinforcement schedule is FR1 or FR256, and the lesioned animals do press more than the normals at FR1.

Our doubts, however, are irrelevant, and Grossman may be correct. There are, however, at least two facts with which Grossman's hypothesis cannot cope. First, it seems likely that an animal with an affective response to food would eat more rather than less often per day, as is the fact. Second, it is simply common sense to expect that an animal with strong "affective responsiveness to all sensory stimuli" will be a very active animal indeed, but the lesioned animal is presumably hypoactive.

Neither the satiety nor the emotionality theories, then, can cope with all of the currently available facts about the lesioned animal. Let us turn next to our own schema relating reactivity to stimulus prominence and examine how well it fits existing animal data, what its shortcomings are, and where it leads.

It seems self-evident that these hypotheses about stimulus sensitivity can explain the emotionality of the lesioned animal and with the exception of meal frequency—a fact to which we shall return—can account for virtually all of the facts about the daily eating habits of these animals. An animal that is hyper-responsive to prominent cues should, when faced with food, eat more of it and eat more rapidly than a less responsive animal. When confronted with a threatening hand or a prodding stick, the hyper-reactive animal should, of course, respond more vigorously. This theory, then, does account for the emotionality of these animals, for many of their daily eating habits, and, consequently, for their obesity.

It should be noted that in each of these instances, the lesioned animal, in contrast to its control, is excessively responsive. There is, however, an array of

[3]Without the special emphasis on emotionality, this is, in many ways, a hypothesis which is similar to our own original notion of the external sensitivity of the obese. As such it suffers from the same defects since, as the reader will see, it cannot account for the numerous instances of hyporeactivity that are characteristic of the lesioned animal.

facts, each indicating in some domain that the lesioned animal is nonresponsive, in fact, virtually torpid. It is well established, for example, that the lesioned rat is hypoactive. This is, of course, the same order of fact as those that compelled the revision of our first notions about human obesity—instances where the obese proved less responsive to food cues than normals. To cope with these negative cases, it was necessary to invent the dimension of stimulus prominence and postulate the interactions specified in Fig. 1. Direct experimental tests strongly support these hypotheses, and for humans, these ideas are, by now, more than *ad hoc* wishful thinking. How well can these same ideas cope with these instances of nonreactivity in the lesioned animal?

To demonstrate that this formulation is as applicable to the VMH-lesioned animal as to the obese human will require the experimental demonstration that the obese animal is more reactive to prominent stimuli than is the intact animal and less reactive to remote stimuli. As a clear instance of this possibility, consider the perverse and fascinating fact that though lesioned animals will outeat normals when food is easily available, they simply won't work for food—an instance of nonresponsivity. In our terms, this may prove only that a remote food stimulus, such as food at the end of an alley, will not evoke a food-acquiring response. If these ideas are correct, it should be anticipated that, in contrast to the results when the food cue is remote, the lesioned animal will work harder than the normal animal when the food stimulus is prominent and compelling. Johnson's experiment (Chapter 8), of course, has demonstrated this pattern of results for humans. As yet, there has been no test of these ideas with the lesioned animal.

Within the existing animal research literature, one can even now, however, find indications that the prevailing notion of the lesioned animal's hypoactivity is a gross oversimplification, for there exist clear instances where the lesioned animal is at least as active, in some cases considerably more active, than the intact animal. Considering first gross activity, as measured in stabilimeter mounted living cages or in running wheels, a variety of studies (Gladfelter & Brobeck, 1962; Hetherington & Ranson, 1942; Teitelbaum, 1957) have reported dramatically less activity for lesioned than for normal rats. With one exception, these studies report data in terms of total activity per unit time, making no distinction between periods when the animal room was quiet and undisturbed and periods involving the mild ferment of animal-tending activities. Gladfelter & Brobeck (1962), however, report activity separately for the "43-hour period when the constant-temperature room was dark and quiet and the rats were undisturbed" and for the "five-hour period when the room was lighted and the rats were cared for [p. 811]." During the quiet time, these investigators find precisely what almost everyone else does—lesioned rats are markedly less active. During the animal-tending period, however, lesioned animals are just about as active as normal animals. In short, when the stimulus field is relatively barren and there is little to react

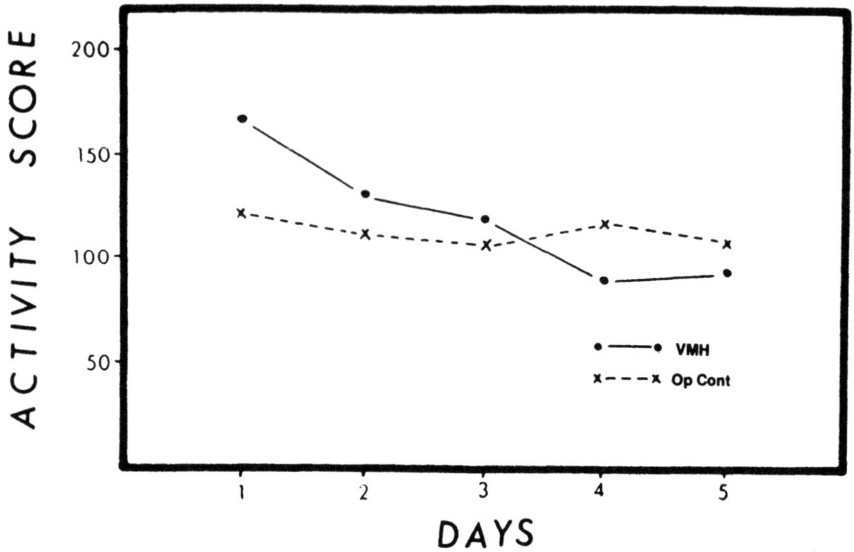

Fig. 2. Mean open-field activity scores of VMH and control groups during five daily activity tests (from Sclafani et al., 1970).

to, the ventromedial animal is inactive; when the field is made up of the routine noises, stirrings, and disturbances involved in tending an animal laboratory, the lesioned animal is just about as active as the normal animal.

Though this is an instructive fact, it hardly proves a hypothesis that specifies that the lesioned animal should be *more* reactive to really prominent cues than the normal animal. Indications that this is the case can be found in Sclafani, Belluzzi, and Grossman's (1970) study of gross movement in the open-field situation. Their procedure involved simply repeatedly placing their animals in an open-field chamber and measuring activity. If we assume that novelty is a major component of the "stimulus prominence" dimension, it should be anticipated that the lesioned rats will be more active than controls on their first exposure to the open field but will become less and less active with repeated exposure. The results of this experiment are reproduced in Fig. 2 where it can be seen that the lesioned animals are significantly more active than operated controls on the first day's test, but that they grow relatively inactive with repeated testing, until by days 4 and 5 they are less active than controls.

Turning next to "activities" other than the gross movements recorded in activity cages or the open-field apparatus, indications are even stronger that under intense stimulation the VMH-lesioned animal is more active than the control animal. Studies of avoidance learning have with fair consistency indicated that the lesioned animal is considerably more reactive than normal animals, for they prove to be better at active avoidance and far worse at passive avoidance.

In active-avoidance studies, the animal must perform some act such as jumping into the nonelectrified compartment of a shuttlebox in order to avoid shock. In the majority of such studies, lesioned animals learn considerably more quickly than do normal controls. In contrast to these results, lesioned animals, in general, do worse than normals in passive-avoidance studies. In these experiments, the animal must inhibit a response to a positive stimulus in order to avoid shock. Most studies find that lesioned subjects either press a lever or touch the food dish more than normals do and accordingly are shocked far more often.

Taken together, these two avoidance situations demonstrate greater reactivity[4] in the lesioned animal. When the situation requires a response if the animal is to avoid shock, the lesioned animal does better than the normal animal. Conversely, if as in approach-avoidance situations, response quiescence is required for the subject to avoid shock, the lesioned animal does far worse than the normal animal. This pair of facts, we suggest, provides support for the contention that there are specifiable circumstances in which lesioned animals will be more active. Obviously only an experiment deliberately designed to test this point can really make the case. Even now, however, it is clear that the prevailing notion that the lesioned animal is hypoactive is a gross oversimplification, for there do exist clear instances where the lesioned animal is considerably more active than the intact animal.

Other Indications of Hyporeactivity

Other than gross activity, there are at least three other behavioral domains in which the drift of the data seems to indicate that the lesioned animal is relatively nonresponsive or hyporeactive. These are sexuality, meal frequency, and the effects of bad taste on food intake.

Sexuality: Though there appears to have been little direct, systematic study of the problem, there is general consensus among workers in the field that the lesioned rat is hyposexual, which if true would certainly contradict a theory which postulates super-reactivity in the face of prominent external stimuli. It is the case, however, that gonadal atrophy, presumably due to inadvertent pituitary disruption, is frequently a consequence of the lesioning operation (Brooks & Lambert, 1946; Hetherington & Ranson, 1940). Possibly, then, sexual activity should be considered a "no-test" situation, artifactually

[4]Reactive, yes, but what about activity in the more primitive sense of simply moving or scrambling about the experimental box? Even in this respect the lesioned animals appear to outmove the normals, for Turner et al. (1967) report that "The experimental groups, both mice and rats, emitted strong escape tendencies prior to the onset of shock and in response to shock. Repeated attempts were made to jump or climb out of the test apparatus. This group showed much more vocalization than the control group .... In contrast to the behavior of the experimental animals, the control animals appeared to become immobilized or to 'freeze' both before and during the shock period. Thus, there was little attempt to escape and little vocalization [p. 242]."

distinct from either gross activity or stimulus-bound activity such as avoidance behavior.

Meal Frequency: Though the fact that obese humans eat less often is no problem, the fact that obese rats also eat less often is awkward, for it's a bit difficult to see how food-stimulus intensity can vary for a caged rat on an ad-lib schedule. Brooks et al. (1946) suggest an amusing possible explanation which could account not only for this fact but for the fact that when unstimulated the hyperphagic rat is so notoriously inactive. They write, "Activity measurements strengthened the conclusion that these rats with lesions tended to fill their stomachs with food, then be inactive and sleep until they could eat again [p. 738]." Though the precise interpretation of this sentence is ambiguous, it does suggest the possibility that the lesioned animals who, as we know, eat considerably larger meals than normals, simply stuff themselves until they become drowsy and fall asleep—a possibility which could account not only for their relative infrequency of eating but for the fact, so troubling to ours as well as all other theories about the obese animal, that they do stop eating.[5]

The Effects of Bad Taste: One final indication of hypoactivity is the lesioned animal's propensity to eat very little bad-tasting food. Though on the face of it, this fact might appear to be compatible with our formulation, it is, given the results of the studies of passive avoidance, oddly disconcerting for our view of matters. In passive-avoidance experiments, a painful event, shock, is temporally coincidental with a pleasant event, such as drinking water or obtaining food. In such circumstances, the lesioned animal is more responsive and is shocked considerably more often than its control. In experiments on taste, an unpleasant adulterant is added to the animal's food and in most such studies, the lesioned animal proves to be considerably less responsive, for it eats far less than do controls. In short, for one kind of unpleasant event, the lesioned animal is more reactive; for another kind of unpleasant event, the lesioned animal is less reactive. Though either fact alone can probably be reconciled with this still loose formulation, the two facts together appear to be flatly incompatible.

We suspect that the key to this dilemma lies in the fact that the passive-avoidance learning setup is a complex stimulus situation with one set of stimuli leading to approach tendencies and the other to avoidance tendencies. On the other hand, the bad taste situation, at least at very high concentrations of the adulterant, might be considered a simple stimulus situation—with no saving graces the food simply tastes overwhelmingly bad. If this analysis is correct, it does suggest that in taste experiments one could create the counterpart of the complex approach-avoidance stimulus pattern characteristic of the avoidance-learning situation by decreasing the

[5]Obviously an externally sensitive organism (or, à la Stellar and Miller, an organism with an excised satiety center) should, in the presence of unlimited food, never stop eating.

concentration of the adulterant so that the bad taste didn't simply overwhelm all of the positive characteristics of the food. Under such circumstances, the lesioned animal should be expected to outeat the normal animal.

There have, in fact, been two experiments which varied the concentration of quinine added to food. Miller et al. (1950) found precisely the anticipated pattern. They describe their results as follows: ". . . the high-fat diet [was] gradually made bitter by progressively adding more quinine to it each day from the 61st to the 70th day after operation. While the concentration of quinine was still relatively weak, the operated animals ate reliably more than the controls, the daily means being 26.4 and 15.1 g, respectively ($p < .01$). Increasing amounts of quinine decreased the intake until, by the 70th day, when the proportion was 1,024 mg of quinine per 100 g, the daily means were down to 2.8 and 8.8 g for the operated and control animals, respectively ($p < .01$) [p. 258]."

Hamilton and Brobeck (1964), working with monkeys, also found that at low concentrations of quinine, operated animals outate normal animals. At their highest concentrations of quinine, the operated and control animals ate roughly the same reduced amounts of food. The failure of these investigators to completely replicate the Miller et al. (1950) findings with rats may simply be due to a failure to employ a sufficiently high concentration of quinine. In any case, it is clear that at relatively low concentrations of quinine the results of the experiments on bad taste do parallel those of the experiments on passive avoidance. When the food is mildly bad and presumably still has some positive qualities, the lesioned animals are indeed more responsive than controls.

To summarize the line of thought that has guided this chapter, the hypothesis that the obese are more responsive to prominent cues and less responsive to remote cues than are normal subjects has for humans proven a fruitful notion. Several experiments directly support the hypothesis, and a wide variety of data concerned with both eating and noneating behavior can be subsumed under this formulation. Because of the persistent and tantalizing analogies between the obese human and the VMH-lesioned rat, we have attempted to extend this hypothesis to what is known about this experimental preparation. Much of the existing data about this animal either fit or can be plausibly reinterpreted to fit these ideas. Obviously, the crucial experiments have yet to be done.

11
DYNAMIC AND STATIC HYPERPHAGIA

In conjunction with our earlier work (Schachter, 1971a), the various studies presented in this monograph have consistently demonstrated an association between obesity and external sensitivity. Though there is the implication in our speculations that externality leads to obesity, we have meticulously avoided consideration of causality, for from the experiments proper there is no more reason to assume that externality causes obesity than that the reverse is true. Indeed, on a common-sense level, one can make a very good case for the proposition that these experiments tell us very little new or interesting but simply spell out the consequences of being fat. After all, the obese animal, human or rat, has an immense reserve of food energy. He may be more finicky and less responsive to internal cues because he can physiologically afford to be so. Similarly, when unstimulated he may move less simply because it's harder to do so. Though such an explanation can by no means account for all of the facts (e.g., emotionality, active avoidance learning, etc.), it does seem such simple good sense that despite its irritating banality, it cannot be dismissed out of hand. Because the matter is, in a sense, crucial to the interpretation of this entire body of data, we will in this chapter digress from the general theme of the monograph in order to consider the implications of our analyses and experiments for the causal question.

There are fortunately experimental preparations which to some degree permit us to make causal inferences. By comparing formerly fat subjects to those who are presently obese we can, to an extent, isolate the contribution of current obesity. Similarly, by comparing dynamic to static VMH-lesioned animals, one can evaluate the effects of the lesion independent of obesity. The static animal is one that has been lesioned and has grown obese. The dynamic animal is one that has been lesioned and is either experimentally studied before it has reached full obesity, or, is a lesioned animal that after reliably demonstrating hyperphagia is reduced to pre-operative weight levels by limiting food consumption.

To compare the two types of lesioned animals, we shall review the results of those studies where both dynamic and static animals were tested in a single study. Because of the problems of comparing studies conducted in different laboratories, employing different procedures, strains of animal, etc., we have deliberately excluded from present consideration those studies in which only dynamic or only static animals were subjects. Though this seems to us a rational and easily defensible decision, it is likely that by limiting our analysis in this way, we overlook other possible comparisons between the dynamic and static animal.

All told, we found a total of eleven studies which compared dynamic and static animals within the framework of a single experiment. In five of these studies, the behavior of the fully obese animal is compared to his own behavior after lesioning but before achieving maximum obesity. In six studies, static, obese animals are compared with an independent group of dynamic animals whose weights have been experimentally maintained at normal levels. There appear to be no differences between the two types of studies.

FINICKINESS

Perhaps the most solidly supported finding about hyperphagic animals is that they overeat greatly when fed good tasting food. Table 1 compares the dynamic and the static animal on this behavior.[1] In this and following tables we have calculated D/N (dynamic/normal) and O/N (static obese/normal) ratios in order to convey the magnitude and direction of the effect. Thus, in the Teitelbaum (1955) study, the dynamic rats ate an average of 26.52 grams and normal rats ate 14.20 grams of this diet yielding a D/N ratio of 1.87. We again adopt a batting-average approach to indicate the extent to which the dynamic and static animals correspond in their behavior. The batting average simply shows the proportion of studies in which the D/N and O/N ratios are in the same direction, indicating that dynamic and static animals behave in the same way. In this particular case, in all four studies both static and dynamic animals outeat normal animals and, as can be seen by comparing the means of the ratios of the individual studies, to much the same extent.

For bad-tasting food, the differences between lesioned animals and normals are not consistent from study to study as can be seen in Table 2. In the Hamilton and Brobeck (1964) study, hyperphagic monkeys, whether static or dynamic, ate considerably more than normals. In the Graff and Stellar (1962) study, the dynamic and static groups both ate less than

[1] Because of space limitations, we have simply listed at the bottom of each table the studies whose data are summarized in the table. The relevant details of each of these studies are described in the Appendix which also specifies the criteria for assigning an animal to the static or dynamic classification and identifies precisely the data on which these analyses are based.

TABLE 1

The Effects of Good Tasting Food on Consumption

Batting average	Mean D/N	Mean O/N
4/4	1.64	1.59
Studies		
Corbit & Stellar (1964)	1.62	1.41
Graff & Stellar (1962)	1.13	1.69
Lipton (1969)	1.93	1.18
Teitelbaum (1955)	1.87	2.06

normals. And in the Teitelbaum (1955) study, dynamic and static animals behave in opposite directions. In part, these differences between studies may be due to different animals and dissimilar test foods. It does seem to us that the status of this particular fact is so suspect that no conclusion can be drawn about the effects of bad taste. However, comparing dynamic and static animals within a study, in two of these three studies the O/N and D/N ratios are in the same direction.

TABLE 2

The Effects of Bad Tasting Food on Consumption

Batting average	Mean D/N	Mean O/N
2/3	1.14	.90
Studies		
Graff & Stellar (1962)	0.23	0.92
Hamilton & Brobeck (1964)	1.66	1.64
Teitelbaum (1955)	1.54	0.14

TABLE 3

Amount of Food Eaten Ad Lib

Batting average	Mean D/N	Mean O/N
6/7	1.62	1.22
Studies		
Brooks et al. (1946)	1.76	1.13
Ferguson and Keesey (1971)	1.44	1.35
Graff and Stellar (1962)	0.94	1.38
Hamilton and Brobeck (1964)	1.63	1.32
Teitelbaum (1955)	1.58	1.25
Teitelbaum and Campbell (1958)—solid diet	1.94	1.07
Teitelbaum and Campbell (1958)—liquid diet	2.04	1.06

EATING HABITS

Examining first the amount eaten daily when food is presented ad-lib, the figures in Table 3 indicate that in six out of seven studies, both dynamic and static animals eat more food than normals, though the effect is considerably stronger for the dynamic than the static groups. In the one study in which this similarity does not hold, Graff and Stellar (1962) found surprisingly that dynamic, not static animals, ate less food than normals.

Considering next the relative number of meals eaten each day, again dynamics and statics tend to behave similarly. As indicated in Table 4, in two

TABLE 4

Number of Meals Per Day

Batting average	Mean D/N	Mean O/N
2/3	1.03	0.85
Studies		
Brooks et al. (1946)	0.90	0.94
Teitelbaum and Campbell (1958)–solid diet	1.21	0.75
Teitelbaum and Campbell (1958)–liquid diet	0.97	0.86

TABLE 5
Amount Eaten Per Meal

Batting average	Mean D/N	Mean O/N
3/3	1.91	1.29
Studies		
Brooks et al. (1946)	2.08	1.17
Teitelbaum and Campbell (1958)–solid diet	1.65	1.50
Teitelbaum and Campbell (1958)–liquid diet	2.00	1.20

out of three studies both groups ate fewer meals than normals. In the third study, dynamics ate more frequently than control animals while statics, again, ate less often. In this study animals were fed pellets, while in the other two a liquid diet or wet mash was provided.

The two lesioned groups are alike in the amount of food eaten per meal. Table 5 shows that in each study both dynamic and static animals eat more food at each meal than do normal weight controls. Again, there is an exaggerated effect for dynamics.

Finally, with regard to speed of eating, the data are less clear. There is only a single study which compares the eating rate of hyperphagic and normal animals, and as indicated in Table 6, the results differ for liquid and solid diets. Dynamic and static animals both drink liquid diet more slowly than normals, but when fed pellets, the static obese eat quite rapidly and the dynamic animals do not. However, other studies (Brobeck, Tepperman, &

TABLE 6
Speed of Eating

Batting average	Mean D/N	Mean O/N
1/2	0.89	1.08
Studies		
Teitelbaum and Campbell (1958)–liquid diet	0.93	0.89
Teitelbaum and Campbell (1958)–solid diet	0.85	1.28

TABLE 7

The Effects of Work on Food Consumption

Batting average	Mean D/N	Mean O/N	Reinforcement schedule
2/2	2.26	1.62	FRI
2/2	0.34	0.10	FR256-1024
Studies			
Hamilton & Brobeck (1964)	2.67	2.00	FR1
Teitelbaum (1957)	1.85	1.23	FR1
Hamilton & Brobeck (1964)	0.50	0.17	FR1024
Teitelbaum (1957)	0.18	0.02	FR256

Long, 1943; Wheatley, 1944) have noted, without reporting data, that dynamics also eat solid foods more quickly than do normals.

For many of their eating habits then, there are strong similarities in the behavior of dynamic and static hyperphagics. Both groups eat more per day and eat more at each meal than do normals. Both eat prodigious amounts of tasty food. Although there are fewer studies of noneating behavior using dynamic and static animals, there are relevant data in at least three areas.

EFFORT

Turning first to the effects of work on food consumption, the figures in Table 7 indicate that both dynamic and static obese animals press far more

TABLE 8

Activity

Batting average	Mean D/N	Mean O/N
1/1	.44	.45
Study		
Teitelbaum (1957)		

TABLE 9
Emotionality

Batting average	Mean D/N	Mean O/N
2/2	5.4	7.3
Study		
Singh (1969)	5.4	7.3
Sclafani (1971)[a]	D > N	O > N

[a]Sclafani reports 0 emotionality for Normal animals, making it impossible to compute ratios. The emotionality scores for his lesioned animals are presented in Table 9 of this volume's Appendix.

often than normals when at FR1 each bar press earns a pellet. However, as more work is required for each pellet, both lesioned groups reduce the number of bar presses they make until finally, at schedules requiring several hundred responses per pellet, they press considerably less often than do normals. It seems clear that dynamic and static hyperphagics respond similarly to changes in schedule.

ACTIVITY

Next we compare the activity levels of the two groups. From the single study in which data for both are reported, the figures presented in Table 8 show that they are virtually identical. Whether or not they are obese, lesioned animals are considerably less active than normals.

EMOTIONALITY

The final datum concerns the emotionality that hyperphagic animals exhibit. When tested for responses to handling, capturing, and other relatively aversive stimulation, both dynamic and static animals appear far more emotional than normals, as indicated in Table 9.

On all noneating behavior for which we have data for both dynamic and static groups in a single study, their responses are strikingly similar. Lesioned animals behave very differently from normals, and the onset of obesity is evidently not necessary for these changes to appear.

All told, then, from the published data in eleven papers, it has been possible to make 29 comparisons of the behavior of dynamic and static obese animals. When compared with normal animals, in 25 of these 29 comparisons, the two groups of lesioned animals behave in a similar fashion. On "amount eaten per meal" and "amount eaten ad-lib" the responses of the dynamic animal is exaggerated and it would appear, as Corbit and Stellar (1964) have maintained, that the degree of obesity does exert an inhibitory influence on consumption of ordinary food. Our analyses support this conclusion as far as food consumption goes but do suggest that this conclusion is limited to this one domain, for on virtually all other dimensions—from finickiness to emotionality and hypoactivity—the two groups of animals are quite similar. It would appear that obesity, *qua* obesity, is not the cause of most of the variety of behaviors associated with the hyperphagic syndrome.

Turning next to humans, it is possible again to evaluate the contribution of current obesity to some of the behaviors that our experiments have demonstrated are associated with obesity. In two of our group's experiments, brief weight-history questionnaires were administered at the end of the study. With the aid of questions such as, "Have you ever been on a diet to lose weight?" it was possible to identify normal subjects who had once been fat. In effect, such subjects are the self-selected counterparts of the dynamic rats that have been fattened up and then reduced to normal weight levels. One of these studies was Nisbett's (1968b) experiment on the effects of taste, good and bad, on the amounts eaten by obese and normal subjects. As we know, all studies have demonstrated that the obese, animal or human, eat more good tasting food than do normals. For bad tasting food, the data are considerably less consistent, but a weak case can be made for the proposition that the obese eat less bad food than do normals.

In Nisbett's experiment, his subjects ate either a very fine and expensive vanilla ice cream or a cheap brand of vanilla with quinine added. Table 10 presents the amounts of ice cream eaten by his currently normal subjects and compares those who were once fat with those who were never fat. Subjects with a history of having been overweight ate considerably more good ice cream and less bad ice cream than subjects who were never fat (interaction

TABLE 10

The Effects of Taste on Currently Normal Subjects with Different Weight Histories (from Nisbett, 1968b)

	Grams of ice cream eaten by normal subjects who were:	
Taste	Never fat	Once fat
Good	136.9	198.3
N	16	12
Bad	38.6	18.1
N	19	9

TABLE 11
The Effects of Work on the Eating Behavior of Normal Subjects with Different Weight Histories

	No. of never fat normals who:		No. of once-fat normals who:	
Nuts have:	Eat	Don't eat	Eat	Don't eat
Shells	10	8	0	2
No Shells	8	9	3	0

$p = .02$). Normal subjects who were once obese behaved as we expect currently obese subjects to behave.[2]

The other experiment in which it is possible to make this comparison is Schachter and Friedman's (Chapter 2) study of the effects of work on eating behavior. In this study, subjects had the opportunity to eat, if they wished, nuts which had shells on them in one condition and no shells in another condition. It will be recalled that this manipulation had a striking effect on the obese, who ate if the nuts had no shells but did not do so if the nuts had shells. For normal subjects, as a group, there was no effect. Table 11 compares the behavior in this study of the normal subjects who had never been fat with that of those who had once been fat. For normals who have never been fat, shells or no shells makes virtually no difference. Though there are relatively few normals who once were fat, without exception these subjects behave precisely as do currently fat subjects. When the nuts have no shells, they eat; when the nuts have shells, they don't.

It does appear that for both men and VMH-lesioned animals, current obesity is not the cause of most of the characteristics that experimentally have been demonstrated to be associated with obesity—a conclusion that goes far in explaining the remarkable recidivism that characterizes the obese condition.

[2]Indeed, in Nisbett's experiment they behave more as we expect obese subjects to behave than do his obese subjects themselves, for in his study (in contrast to Decke, 1971) his currently obese subjects do not eat less of bad tasting food than do his currently normal subjects. We admit to being as puzzled by this fact as we are by the generally confusing and inconsistent set of studies on the effects of bad taste, whether for animals or men.

PART III
EXTERNAL SENSITIVITY IN NON-FEEDING SITUATIONS

That more than aberrant eating behavior is involved in the obesity syndrome seems by now clear, for obese animals and men seem also to differ from their normal counterparts in emotionality, avoidance behavior, and activity. Our guesses about differential stimulus sensitivity are, of course, an attempt to invent a conceptually economical dimension to integrate the variety of behaviors on which obese and normal subjects appear to differ. A possible implication of this theoretical attempt is that the various findings about eating behavior are simply a special case of a much broader phenomenon. The evidence supporting such an interpretation is, at the moment, suggestive but hardly conclusive, for it is easily conceivable that differences in emotionality and activity directly derive from obesity rather than represent independent manifestations of a more broadly based pattern of differences. Though the comparisons of dynamic and static animals drawn in Chapter 11 make this an unlikely causal chain for animals, it is certainly a plausible interpretation of the human data. No exotic argument is required to defend the suggestion that obese people are emotionally arousable and volatile simply because they are unhappy people rather than because of stimulus sensitivity.

Even the beginnings of a case for the generalized stimulus sensitivity notion demand the demonstration that there are differences between obese and normal people that are simply unlikely to be consequences of obesity. The search for such differences has been of sporadic concern to our group, and our early efforts (Schachter, 1971a) yielded conflicting and largely negative results. In one line of investigation, we investigated the relationship of obesity to field dependence. Generalizing from their experiments on perceptual style to a wide realm of personality characteristics, Witkin, Lewis, Hertzman, Machover, Meissner, and Wapner (1954) have identified a dimension they call field dependence, by which they refer to the extent to which an individual relies on field or, in the present sense, external cues. Since our original notions on the external versus internal control of eating behavior bore a striking resemblance to this more general formulation, it seemed a fair guess that our results on eating behavior were a special case of field dependence. Supporting this guess, Karp and Pardes (1965) had found that the obese proved more field dependent than normals as measured by performance on the embedded figures test. In an attempt to replicate these findings, two independent investigations (conducted by Pliner and Kay and by Maher, Mayhew, and Zellner—both reported in Schachter, 1971a) were carried out in our laboratory. There were no differences on the embedded figures test between obese and normal subjects—all Columbia college students. Since these subjects were drawn from the population that had served as subjects in most of our eating experiments, it seems unlikely that the obese-normal differences consistently found in these studies can be considered a special case of differential field dependence.

In another line of inquiry, Maher, Mayhew, and Zellner (reported in Schachter, 1971a) attempted to extend the external-internal control schema to the states of thirst and urination anticipating that, as with eating behavior, the obese would prove more vulnerable to external cues than would normal subjects. There were no differences between the two groups. This experiment was conducted before we had become aware of the importance of the cue prominence dimension. Following Ross' demonstration (Chapter 7) of the crucial importance of this dimension, Kozlowski and Schachter, in a still unpublished study, deliberately manipulated the salience of the water cue and did indeed find that obese subjects drank much more water than normals when the water cue was prominent and about the same amount when the cue was remote. The behavior involved was still a consummatory behavior, but this was our first real indication that this set of ideas might apply to more than just the consumption of food.

The following set of papers present experimental efforts to test the notion of generalized stimulus sensitivity and its implications in a variety of contexts unrelated to consummatory behavior. The paper by Rodin, Herman, and Schachter (Chapter 12) is a first effort to simply search for differences between normals and the obese on a variety of measures, such as reaction time and tachistoscopic recognition, that have nothing to do with food or drink but do presumably tell us *something* about stimulus sensitivity. This is an openly shotgun effort, and we make no pretense that any systematic line of thought guided our choice of measures. It does prove to be the case that the obese are simply better at discriminative reaction time, immediate recall, and tachistoscopic recognition. Despite the fact that these results are statistically reliable and, in fact, have been replicated several times, we present this first paper with considerable uneasiness. We think these data should be in the literature, but we are not yet convinced that they mean what we suggest they mean. They may tell us something about differential stimulus sensitivity; they may also tell us something about differential motivation. Whether we like it or not, there is general consensus on what constitutes good performance on the sort of measures employed in this study. To be fast is better than to be slow; to remember many items is better than to remember few. It is conceivable that in selecting subjects according to degree of obesity we are also inadvertently selecting subjects who are differentially eager to please and to make a good impression. Fat subjects, plagued by the psychology of the minor pariah, may simply be driven to impress the experimenter more than are normal subjects.

In part, because of this major ambiguity, the remaining experiments presented in this section were all deliberately designed to test interaction predictions—to specify the conditions under which fat subjects should do worse than normals as well as the conditions in which they should do better than normals. In addition, several of these experiments are concerned with variables—such as time perception—for which it is particularly unlikely that an explanation in terms of differential motivation or desire to please could possibly explain the experimental results.

These several experiments were all designed to test, in non-eating situations, various implications of the hypothesis developed in Chapter 10, that the relationship of cue prominence to reactivity is considerably stronger for obese than for normal subjects. To test this hypothesis, Rodin in Chapter 13 examines the effects of distraction on performance. She reasons that if the hypothesis is correct, highly prominent, distracting stimuli should be more disruptive for obese than for normal subjects when they are performing a task requiring concentration. Her experiment indicates that this expectation is correct.

Finally, Pliner in Chapter 14 describes a series of experiments which examine (*a*) the differential effects of cue prominence on the time perception of obese and normal subjects, (*b*) the differential effects of cue prominence on ideation and on pain perception, and, (*c*) the generalizability of the findings on emotionality to positive as well as negative emotions. These experiments all indicate that the stimulus binding hypotheses do generalize to a variety of behaviors having nothing at all to do with eating.

12
OBESITY AND VARIOUS TESTS OF EXTERNAL SENSITIVITY

Judith Rodin, C. Peter Herman,
and Stanley Schachter

That a prominent cue is more likely to trigger a food acquiring response in an obese than in a normal subject appears by now to be a clear fact. Its interpretation, however, is not, for there is no particular reason, as yet, to think of this fact in either motivational or perceptual or information-processing terms. Indeed even the simple parameters of the fact are obscure, for we are not as yet sure, for example, if this fact is specific to food cues or indicative of a general reactivity to prominent cues.

Given such ambiguities, our strategy at one stage in these investigations was simply to gather as many possibly relevant facts as we could in the hope that the sheer accumulation of information would eventually force a more conceptually precise picture of the phenomenon. In a series of small studies we simply compared the performance of obese and normal subjects on a variety of tests which, in the literature of experimental psychology, have been assumed to indicate *something* about the likelihood that a stimulus will trigger a response. We reasoned that if the obese are generally more responsive to salient stimuli than are normals, they should react more quickly to external cues, take in more environmental stimuli, possibly remember them better, and so forth. To test such guesses, we measured the reaction time latencies of obese and normal subjects, their immediate recall for items presented briefly on a slide, and their threshold for the recognition of tachistoscopically presented stimulus material.

METHOD

Three different experiments were designed using several tests of "external sensitivity." The first experiment measured, in sequence, simple and choice reaction time, immediate recall, and finally the test of shock sensitivity described in Rodin, Elman, and Schachter (Chapter 3). The second experiment again compared obese and normal subjects on simple and disjunctive reaction time. Experiment 3 measured tachistoscopic recognition thresholds.

In all three studies, subjects heard a brief introduction which explained that the purpose of the experiment was to measure reactions to various sensory stimuli. They were told that the tests would "provide information on how individuals encode and respond to their sensory environment . . ." Depending upon the experiment, subjects were then tested on one or more of the following measures:

Reaction Time

The subject sat at a table, facing a partition which displayed two 6.3 volt bulbs spaced about 18 inches apart, at eye level. Directly beneath each bulb was a telegraph key. The reaction time apparatus was identical to that described by Rodin (Chapter 13).

In simple reaction time measures, the subject was instructed to use the right key if he was right handed or the left, if left handed. A single light, ipsilateral to the dominant hand, was used as the stimulus. He was told to press the key with his forefinger at the ready signal, and to release it as soon as he saw the light appear. The subject was instructed to rest the heel of his hand on the table in front of the key, and to respond by raising his finger rather than by withdrawing his whole hand. Intervals between the ready signal and the light onset were randomized, ranging from 4-9 seconds. Eight-second intervals were given between trials. There were 2 practice trials followed by 10 experimental trials.

In choice reaction time, the subject responded to the ready signal by pressing both keys. When a light appeared, he responded correctly by releasing the key controlled by the contralateral hand. In other words, if the left light appeared, he released the right key. The exact procedure has been detailed elsewhere (Rodin, Chapter 13). Subjects were given 2 practice trials and 10 experimental trials.

Immediate Recall

The subject was seated in a darkened room about 10 feet from a white screen. He was told that a group of words or numbers or objects would appear briefly on the screen and after each presentation he would be asked to recite aloud all the items he could remember. An experimental slide was then presented for 5 seconds, followed by a blank slide for 15 seconds during which time the subject reported the items.[1] At the end of 15 seconds, there was a ready signal and a new slide was projected.

[1]Subjects were told to take all the time they needed to report the stimuli and all finished well before the 15-second intertrial interval was completed.

Eight slides were presented for 5 seconds each, separated by 15-second periods. A practice slide was always given first and the seven experimental slides were presented in a different, randomly-generated order for each subject. All slides contained 13 items arranged in 3 columns. The content of the slides is described below:

Practice slide: 13 two-letter, non-food-related words
Experimental slides: 13 three-letter, non-food-related words
13 four-letter, food-related words
13 two digit numbers
13 non-food-related objects (2 slides)
13 food-related objects (2 slides)

Tachistoscopic Recognition Thresholds

The subject was asked to identify a word which was flashed briefly in the visual field of the tachistoscope. Increasingly long flashes were presented for each word, and he was instructed to state after each flash what he thought he had seen. The subject was told that he was permitted to guess or to say he had no idea what the word was. When he identified the word correctly, a new word was presented. He was instructed to rest his head against the eye piece of the tachistoscope and to keep looking into the apparatus. In this way, the subject was light adapted to a fairly constant level after each exposure. A rest period was given every few words.

The words were exposed in a 2-channel tachistoscope. One field was blank and the stimulus field remained lighted between word presentations. The words were typed, in capital letters, on $2'' \times 5''$ index cards and appeared in the center of the stimulus field about 20 inches from the subject's eyes.

A set of 25 words selected from the Thorndike and Lorge (1944) word frequency count were presented to each subject. Four words served as practice words, and the remaining 21 were presented in a different, randomized sequence for each subject in order to control for practice effects.[2]

Following the ascending method of limits, the flash duration was systematically lengthened until the exposure word was accurately reported. The duration for the first presentation of each word was set at 5 msec. and was increased by 5 msec. until a 30 msec. duration was reached, after which exposures increased in increments of 10 msec. Each stimulus word was presented until the subject correctly identified that word on two successive trials.

Subjects

Twenty obese and 20 normal-weight Columbia males participated in experiment 1. Obese subjects ranged from 17.2 to 42.2% overweight and nor-

[2]The test words were: book, cargo, cat, deer, desk, elbow, ferry, fuse, horse, house, knot, mile, motor, nail, pulp, sack, table, vast, wood, word, year.

TABLE 1
Mean Reaction Time (Msec.) in Experiment 1

Subjects	Simple RT	Choice RT
Normal	197.8	424.0
Obese	212.0	393.4
Obese vs. Normal	$t = 2.84, p < .01$	$t = 2.07, p < .05$

mals from −7.9 to 9.0%. Ten obese subjects, ranging from 15.4 to 30.0% overweight, and 10 normals with a weight range of −9.4 to 6.4% overweight, served in experiment 2. Twenty obese and 20 normal New York University males participated in experiment 3. Their weights ranged from 15.7 to 57.6% overweight for the obese group and −7.4 to 10.1% overweight for the normals.

RESULTS

Reaction Time

A score for each subject was based on his median response latency for the 10 experimental trials. The mean group scores are presented in Table 1 where it is evident that overweight subjects were slower than normals at making a simple response to a single light stimulus. On the other hand, they responded more quickly than normals in a choice RT measure. Since there was no *a priori* reason for anticipating this particular pattern of results we replicated the study.

In experiment 2, all subjects were tested on both simple and choice reaction time; however, half the subjects were tested first on simple and then on choice RT, and half were tested in the reverse order. Since simple reaction time always preceded the choice reation-time measure in the first experiment, it seemed possible that simply due to negative transfer, those subjects who were better at the simple response—the normals—would respond more slowly in the choice task.[3] As indicated in Table 2, obese and normal weight subjects were not differentially affected by order of presentation, and the same pattern of between-group differences characterizes both sequences.

Combining both orders of presentation, the simple and complex reaction-time pooled means are consistent with the data obtained in the first ex-

[3]The choice RT measure required subjects to respond with a different hand to the identical light used in the simple task. Subjects who responded to a right light by releasing the right key in simple RT had to release the left key for a correct response in the disjunctive RT.

TABLE 2

Mean Reaction Time (Msec.) for Replication (Experiment 2)

Subjects	Order of presentation			
	Simple-choice sequence		Choice-simple sequence	
	Simple RT	Choice RT	Simple RT	Choice RT
Normal	201.7	439.0	234.0	442.3
Obese	223.0	396.7	238.0	368.2

	Means combined across order of presentation	
	Simple RT	Choice RT
Normal	217.9	440.7
Obese	230.5	382.5
Obese vs. Normal	$t < 1$, n.s.	$t = 3.53, p < .01$

periment. Obese subjects were somewhat, although not significantly, slower than normals in simple reaction time and considerably faster in disjunctive reaction time. Smith and Boyarsky (1943) also report some tendency for heavier subjects to respond more slowly than light subjects to a single loud sound stimulus. The strong differences in choice reaction time have now been replicated a third time (Rodin, Chapter 13).

Immediate Recall

After each of 7 slides, subjects were asked to recall aloud every item they remembered from the slide. Out of 13 possible items on each slide, obese subjects remembered, on the average 6.52 items correctly while normals recalled 5.76, a difference significant at better than the .002 level. In addition, as indicated in Table 3, overweight subjects made somewhat fewer incorrect responses than normals. It is unlikely then that they were better at recalling the slides simply because they indiscriminately gave more responses.

Four of the slides contained non-food-relevant stimuli while three showed pictures of food or listed food words such as lamb, veal, etc. Since overweight subjects are more responsive to food-relevant cues, it is possible that the differences in overall recall may be accounted for largely by differences on the

TABLE 3
Mean Number of Slide Items
Correctly and Incorrectly Recalled

Subjects	Mean correct/slide	Mean errors/slide	Percentage of total answers incorrect
Normals	5.76	.41	6.6
Obese	6.52	.26	3.9
Obese vs. Normal	$t = 3.34, p < .002$	n.s.	

food slides. When the mean recall scores were divided for food and non-food-relevant slides, obese subjects still recalled significantly more items than did normals ($t = 2.78, p<.01$) on slides unrelated to food. The mean scores were 6.41 for the obese and 5.53 for normals. Overweight subjects simply did better whether the slides presented food relevant stimuli or not.

A Replication

It was possible to retest these findings on immediate recall within the context of another study of obesity whose purpose is irrelevant to present concerns. This study was conducted at an Eastern girls' college by Elizabeth Decke, Alice Gold, and Katherine Porikos. As a first step in this study, these investigators administered three slides (each, like ours, containing 13 objects or words) following a procedure similar to that employed in our study.

In addition to testing immediate recall, they were able, because of certain peculiar features of the student body of this college, to check an observation that we have made repeatedly in the past. For no apparent reason, among the girls in this college there were a goodly number of exceptionally fat girls—so fat that one is tempted to think of them in terms of metabolic or glandular disorders. In our experiments in the past, we have repeatedly noted that such subjects do not seem to behave in the same fashion as do routinely fat subjects but resemble normal subjects on our various tests of externality. Since in our past studies, we have never had more than one or two such subjects in any particular study, we have been unable, until now, to pursue these first indications of a nonmonotonic relationship.

The performance of normal, obese, and super-obese subjects on these tests of immediate recall is presented in Table 4. It can be seen first that, as in the

TABLE 4

The Effect of Degree of Obesity on Immediate Recall

Subjects	*N*	Mean overweight	Range of overweight	Average no. of items correctly recalled
Normal	35	− 1.1	− 9.7 to + 8.6	5.14
Obese	27	+31.4	+16.0 to +48.6	5.54
Super-obese	8	+71.4	+53.3 to +97.7	4.63

Comparison	*t*	*p* value
Obese vs. Normal	1.95	$< .06$
Obese vs. Super-obese	3.05	$< .01$
Normal vs. Super-obese	1.88	$< .10$

original study, the routinely obese recalled a somewhat greater number of correct items than did normal subjects. In contrast, the super-obese did markedly worse at this task than either routinely obese or normal subjects. In conjunction with our earlier observations, these results do suggest that our various hypotheses about the external sensitivity of the obese human being do not apply to the extreme cases.

Tachistoscopic Recognition Threshold

The exposure duration at which a subject correctly identified a word, followed by correctly identifying it again on the next trial, was considered his recognition threshold. For each subject a mean threshold was obtained for the combined 21 experimental words. The data reveal a significant tendency for the obese to recognize the words at shorter exposures than do normals ($U = 126, p < .05$). The mean tachistoscopic recognition threshold for overweight subjects was 32.8 msec. while normals scored 63.0 msec. on the average. In addition, the variability within the normal group was almost 10 times greater than in the obese. Some normals recognized the words very quickly while others did extremely poorly, often failing to identify correctly a word at 200 and 300 msec. exposures.

DISCUSSION

In three of the four tasks used in these experiments, obese subjects did better than normals. They reacted more quickly when they were required to make a discriminative choice before responding, they recalled more items on slides they had seen briefly, and they recognized words at shorter exposure durations.

Given the caveat, in the introduction to Part III of this monograph, about alternative interpretations of these data[4] we are reluctant as yet to do more than briefly spell out some of the dimensions along which, these data suggest, the obese and normal groups may differ. It may be that, for some reason, the obese are more efficient information processors than normals. They may actually take in more information than normals, due either to sensory or attentional differences, encode it more effectively, or store more in short-term memory. Differences between subjects at any of these steps could have produced many of the reported effects and all would basically support the assertion that for the obese, attention and response seem more directly wired to external stimuli.

Parenthetically, even the slower simple reaction time may actually support this contention. Woodworth and Schlosberg (1954) suggest that concentration on the stimulus to be received (rather than the response to be made) shortens response latency on choice reaction time while it slows simple reaction time. This interaction, which precisely describes our findings for overweight subjects, supports the view that the obese are generally more attentive and responsive to external stimuli.

[4]There is one persistent fact which argues against interpreting this body of data as simply demonstrating the desire of the obese to make a good impression. In two separate studies, the obese are somewhat slower than normals in tests of simple RT.

13
EFFECTS OF DISTRACTION ON THE PERFORMANCE OF OBESE AND NORMAL SUBJECTS[1]

Judith Rodin

In conjunction with the findings on emotionality, activity, and avoidance behavior, the results of the Rodin, Herman, and Schachter (Chapter 12) series of studies provide further evidence that the differences between obese and normal people do extend beyond the eating domain. In an attempt to supply coherence, this melange of apparently unrelated findings has been tentatively interpreted in terms of differential generalized external sensitivity—a still vague notion sometimes called the stimulus-binding hypothesis, which suggests that in some fashion external cues compel the attention of the obese more than of the normal subject.

If these loose ideas have validity, and if they relate to our formulation of the relation of cue prominence to reactivity, it should be expected that in any circumstance, food relevant or not, a prominent external cue will be more likely to compel the attention of the obese. If this is correct, it should follow that the obese will be disrupted in any activity requiring concentration when a distracting stimulus is prominent. On the other hand, if the distracting cue is weak, the obese should be unaffected by it. In fact, if Ross's (Chapter 7) results with food generalize, it could be expected that a weak cue will have less effect on an obese than on a normal subject. In addition to investigating these implications of the stimulus-binding hypothesis, the present study examined two other questions concerning this hypothesis.

First, we considered the duration of a distracting cue's effects. The strong form of the stimulus-binding hypothesis suggests that when a cue is present, the obese respond to it; when it is not immediate, it does not affect them. In

[1] This report is based upon a dissertation submitted to Columbia University in partial fulfillment of the requirements for the Ph.D. degree. The author is greatly indebted to Stanley Schachter for his support in all phases of the study.

other words, it may be that for overweight subjects "out of sight" *is* "out of mind." Consequently, the same irrelevant cues which are compelling and distracting for the obese while they work on a concentration task should not be at all distracting if they precede the task. Presumably, a true stimulus-bound individual would not continue to think about them in the latter case.

Second, an attempt was made to test the generality of the basic stimulus-binding hypothesis. As elaborated above, the hypothesis implies that distracting cues will disrupt productivity in *any* kind of task. Although no one is really ever able to fully test generalizability, the hypothesis was, at least, tested on two totally different tasks—a manual monitoring task involving hand-eye coordination and the minor "mental" activity involved in correcting proof.

METHOD

Overview of Experimental Design and Procedure

Subjects worked on 2 tasks—reaction time and proofreading. During both tasks, distracting stimuli, irrelevant to the task, were introduced for some subjects while others worked without distraction. Tape recordings intended to vary in interest were used as distractors. In addition, some subjects heard tapes during the reaction time and proofreading measures; others listened to tapes prior to these tasks. Twelve experimental conditions—6 groups of obese subjects and 6 normal weight groups—were designed to test the hypotheses outlined above. There were 10 subjects in each group.

Subjects

Sixty obese and 60 normal-weight males recruited from introductory psychology classes at Columbia University participated in the experiment. The weight deviation of the obese group ranged from 15.5% overweight to 62.8% while normals ranged from −9.8% to +9.8% overweight, as calculated from the Metropolitan Life Insurance norms (1959). Obese subjects were, on the average, 44.4 pounds heavier than their normal-weight counterparts.

Procedure

Subjects were recruited for participation in a study on the "physiological and psychological effects of sensory overload." When a subject arrived in the laboratory, he was seated at a table in a small experimental room. On the table were two telegraph keys and an upright panel on which two 6.3-volt bulbs were mounted. The experiment was "explained" as one in a series of studies aimed at determining the effects of multiple sensory stimulation—that is, the stimulation of several sense organs simultaneously—in order to investigate problems of noise pollution. Stress was placed on physiological measures as the primary source of data, and the subject was told that his heart rate would be monitored during the entire period in order to measure physiological responses to the various experimental manipulations. Electrodes were attached to the subject's wrists and he was told that the baseline heart-rate measures would be taken.

After this introduction, a uniform sequence of events followed for all subjects:

1. First experimental session.
   a. Baseline recording period: heart rate measurement.
   b. Reaction-time test.
2. Questionnaires: questions about attention to and difficulty with the various tasks and perceived physiological and emotional state.
3. Second experimental session.
   a. Baseline recording period: heart rate measurement.
   b. Proofreading test.
4. Questionnaires and experimental debriefing.

To test the experimental hypotheses with two different concentration tasks, all subjects worked on reaction time in session 1 and proofreading in the second session. For reaction time measures, the subject was instructed to keep his hands on the table and his forefingers pressed down on both keys. Upon seeing the right light, he was to release the left key, and when seeing the left light, to release the right key. In other words, he had to release the key opposite to the light which appeared in order to make the correct response and stop the timer. When the light went off, he was instructed to put his fingers back on the keys and get ready for the next trial. There were 48 stimulus presentations, 24 for each bulb, and the intertrial interval varied from 4–9 seconds. Reaction times were measured with an electronic timer which was activated by 1000-per-second pulses from a tuning fork. Errors were recorded automatically by an event recorder. The reaction time test period lasted 10 minutes.

In the second session, all subjects proofread a passage for errors. They were told to underline each error and place a check "in the margin next to the line where the error appears." The selection dealt with slums and was taken from Jane Jacobs' *The Death and Life of Great American Cities.* The tape recorder was started whether subjects heard a tape or not, and exactly 10 minutes after beginning, a taped voice announced that the session was completed. Subjects were told to remove their headphones and underline the last sentence read.

To summarize, all subjects worked on both reaction time and proofreading. Each of these tasks was preceded by a presumed baseline heart-rate measurement period which was, in fact, used to manipulate the immediacy of the distracting cues.

Manipulation of Immediacy. To test the prediction that stimulus-bound individuals are only absorbed by immediate stimuli, the distracting tapes were presented *before* the task for some subjects. During the baseline measurement period, these subjects, designated as *PreTask Distraction* groups, were instructed to "pay close attention to the tape and think only about those things suggested on it. In this way we will be certain of controlling your level of attention and concentration during our baseline physiological measures . . ." After hearing the tape, they began working on the reaction time task in the first session or proofreading in the second. No tape played while they were performing these tasks.

In contrast, subjects assigned to the *Distraction* condition always heard a tape while they worked on reaction time or proofreading. These subjects were instructed that "your primary task is to attend to the visual monitoring, but at the same time you will be listening to auditory material so that we may uniformly control each subject's stimulus input . . ." In the baseline heart-rate session which preceded each task, these subjects simply added columns of numbers for 10 minutes.

In summary, there were two groups of subjects who heard the tapes at different times. Some listened to tapes before the reaction time or proofreading measures and then worked on the tasks with no tapes, while the others listened to the tapes and did the tasks at the same time. A third group served as *No Distraction* controls. These subjects heard no tape during either baseline or actual task periods. Regardless of condition, all subjects wore headphones during the ten-minute test sessions.

Manipulation of Interest. In order to manipulate the potency of the competing stimuli, the distracting tapes varied in interest. As a presumably uninteresting distractor, one tape simply recited *random numbers* for 10 minutes. A moderate level of interest was aroused by *Neutral* distraction tapes which in the reaction time session depicted scenes of rain and snow, and in the proofreading period, described sea shells. The rain tape, for example, told subjects to:

"Picture a heavy rainstorm. Picture the sky getting blacker and blacker as the cracks of thunder and bolts of lightning begin . . . picture the first large raindrops that fall . . ."

Similarly, the sea shells tape intoned:

"Consider the smell of sea shells—the raw, briny smell of the ocean. Imagine the salty waves slapping against the shells . . . think of the roar of the ocean when you hold a large shell to your ear . . ."

In contrast, a third pair of tapes describing the atomic bombing of Hiroshima and one's own death from leukemia was expected to be highly interesting and evocative for most subjects. The Hiroshima tape was always used in session 1 and the leukemia tape in the second session. These *Emotional* distraction tapes explicitly described compelling, affect-laden events. In the Hiroshima tape, the burns and mutilation which resulted from the bombing, the ruins and terror, and the dreadful aftermath were all depicted. For example, subjects were told to:

"Picture people whose faces were wholly burned, their eyesockets hollow, the fluid from their melted eyes running down their cheeks . . ."

The leukemia tape described the insidious onset of the disease and its symptoms, how the subject might respond psychologically to having leukemia, how people might fear and ignore him, and how death would occur. For example,

"Consider how the impact of your disease would . . . affect your every significant relationship . . . who would feel distressed and who would feel put upon . . . who would be glad to see you dying in great pain and suffering . . ."

These quotes from the high potency condition undoubtedly make it clear that emotionality as well as high interest was manipulated. This was deliberate. Obviously, to test the distraction hypothesis in its purest form, interest alone should simply have been manipulated. We assumed, however, that this experimental requirement would be satisfied by the random-numbers and neutral-tape conditions. Consequently, given the general theme of this monograph—the investigation of obese human and animal parallels—the high interest condition was added to also test the emotionality of the obese person.

Since there were two experimental sessions—reaction time and proofreading—for each subject, tapes with the same level of emotional arousal were used in both sessions. This procedure was followed in order to avoid possible contaminating effects of one type of tape on another. For example, a subject continuing to think about the bombing of Hiroshima might still be profoundly disturbed in the second session when he was supposedly rather bored from listening to a nondisturbing tape of random numbers. Instead, all subjects hearing about Hiroshima in the first session heard another emotional tape—a description of leukemia—in the second.

While subjects heard tapes of the same emotion and interest level in both sessions, a partial factorial design was employed in which the time of presentation of the taped material was varied. Subjects who listened to a tape before the reaction time, proofread while listening to a tape; subjects who heard a distraction tape during the reaction time task, heard a tape in the baseline measurement period before correcting proof. Finally, subjects who worked without distraction in the first session listened to random numbers in the second, while those who first heard random numbers next worked without distraction.

RESULTS

Potency of Tapes

Cue potency was manipulated by varying how interesting the different tapes were. More interesting tapes were expected to elicit greater attention than boring monotonous tapes. To see whether subjects did, in fact, find the

TABLE 1

Mean Interest Ratings of Tapes[a]

Subjects	Proofreading (Session 2)[b]		
	Numbers	Neutral tape (seashells)	Emotional tape (leukemia)
Normal	9	29.5	51
Obese	6	39	56.5
	Reaction Time (Session 1)		
	Numbers	Neutral tape (Rain)	Emotional tape (Hiroshima)
Normal	23	52.5	61.7
Obese	22	61.5	64

[a]Ratings were made on a scale from 0-100.

[b]The overall lowered interest scores for tapes during proofreading probably reflect the fact that subjects were answering the identical questionnaires a second time and simply gave lower ratings on all questions, not just those concerned with the tapes.

distracting material differentially interesting, they were asked to rate the tapes at the end of each session. From the data reported in Table 1, it appears that the manipulation of interest was successful. In all cases, subjects found the recital of random numbers less interesting than the neutral tapes, which, in turn, were rated less interesting than the emotional tapes. Since the tapes can be ordered along a dimension of interest or compellingness, if our predictions are correct, obese subjects should be relatively unresponsive to the random numbers and highly responsive to, and thus distracted by, the emotional tapes.

Proofreading

Distraction. To test the distraction hypothesis, measures were taken of the number of pages a subject read and how accurately he corrected proof. The first index gives information about the quantity of performance; the other, about the quality.

a. *Number of pages:* This measure reflects the subject's own proofreading speed, since the instructions did not specify how quickly subjects were expected to work nor whether they would have time to finish. The index was ob-

TABLE 2

Mean Proofreading Accuracy
Number Correct/Number Possible on Pages Read

	No distraction	Numbers distraction	Neutral distraction	Emotional distraction
Normals	.471	.505	.514	.568
Obese	.612	.556	.469	.434

Comparison

No Distraction: Obese vs. Normal $t = 2.62^a$, $p < .02$
Emotional Distraction: Obese vs. Normal $t = 2.49$, $p < .05$.
Obese: No Distraction vs. Neutral Distraction $t = 2.66$, $p < .02$.
Obese: No Distraction vs. Emotional Distraction $t = 3.31$, $p < .01$.
All similar comparisons for normal weight subjects were nonsignificant.

[a] All t-tests used a variance based on the mean square error term for the "distraction" analysis of variance.

tained simply by recording on which page, and where on that page, a subject underlined a sentence to indicate the last line read before the session was terminated. Subjects in all conditions read approximately the same number of pages, and statistical analyses comparing obese and normal subjects for each distraction showed no significant effects.

Since neither weight group read more slowly when distracting stimuli were introduced, quantity does not appear to be the crucial variable. However, if the obese continued to read at the same pace but were, in fact, distracted, the quality of their performance should have been seriously impaired. The accuracy index provides these data.

b. *Accuracy:* As a measure of how accurately each subject read proof, his total number of errors was divided by the number of possible errors on the pages he read. Thus, if the subject identified 45 errors out of the possible 90 that he covered, his accuracy index was .50. The group means are presented in Table 2 and plotted in Fig. 1. It is clear that the distraction manipulations had a strong negative effect on the accuracy of obese subjects, while they appear to have had a somewhat facilitative effect on normals (Weight + Distraction $F = 4.81$, $df = 3/72$, $p < .01$). When there were no competing stimuli, overweight subjects were significantly more accurate than

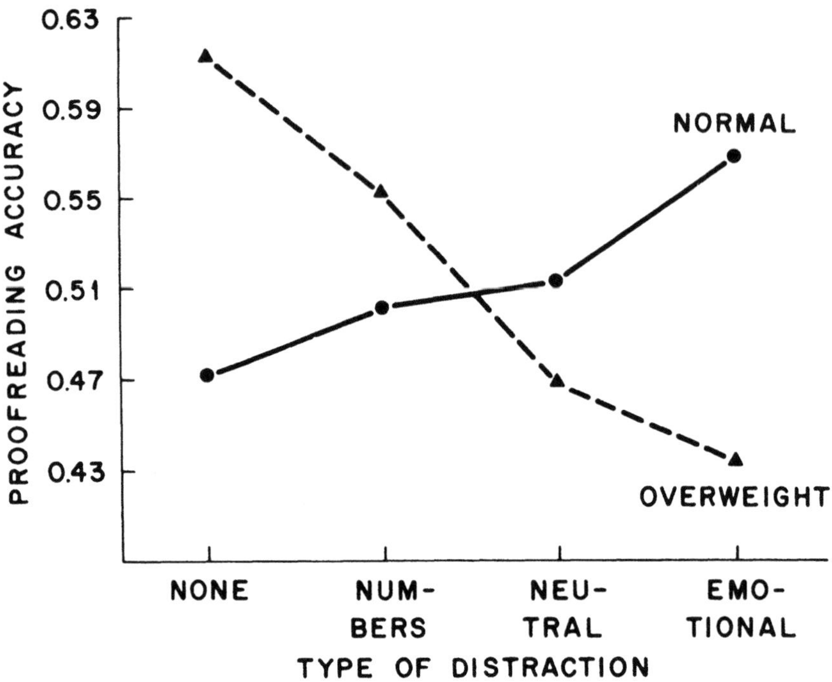

Fig. 1. Accuracy while proofreading.

were normals ($p < .02$). In striking contrast, the obese were significantly more distracted than were normals by the potent arousing tape ($p < .05$). The analysis of variance interaction is significant with $p < .01$, 99% of the effect being accounted for by a linear × linear trend. This is evident in Fig. 1, which shows a linear increase for normals and a linear decrease for the obese as a function of how interesting the tapes were.

It should be specifically noted in Table 2 that in the Neutral Distraction, as well as in the Emotional Distraction groups, obese subjects performed significantly more poorly than in the No Distraction condition. This finding is particularly important to the overall interpretation of the data, since our manipulation did to some extent covary interest and emotionality. To further consider the effects of the interesting, nonemotional tape, obese and normal subjects in the No Distraction and Neutral Distraction conditions were compared. The Weight × Treatment interaction was significant with $F = 4.92$, $df = 1/36$, and $p < .05$. It is clear from these findings for neutral distraction that the disruptive effects of the distracting tape were not necessarily due to their emotional component.

In addition to performance measures, there is independent evidence of the degree of emotional arousal of these subjects. When asked to rate themselves on several indices of emotionality, obese subjects hearing the neutral tape reported a mean emotionality score of 9.5 which was quite similar to the 13.3

TABLE 3
Mean Proofreading Accuracy

Subjects	No distraction	Pre-task neutral distraction	Pre-task emotional distraction
Normals	.471	.598	.555
Obese	.612	.555	.666

given by obese subjects working without distraction ($t < 1$). In contrast, when hearing the highly arousing tapes, overweight subjects reported a mean emotionality score of 29.3 ($t = 2.71$, $p < .01$). Clearly, then, obese subjects hearing the neutral tapes were hardly aroused at all, but their performance was significantly disrupted by the tapes. While a cue's emotional evocativeness undoubtedly contributes to its potency, emotionality alone did not produce the distraction effect.

Effects of Hearing a Tape Before Proofreading. Four experimental groups heard a neutral or emotional tape before correcting proof and nothing during the task. This was intended to determine whether distracting stimuli must be immediately impinging to disrupt performance. As with subjects who listened to the tapes while proofreading, these subjects did not differ in the number of pages read. However, in addition, there were no significant accuracy differences between Pre-task Distraction conditions and the No Distraction baseline (Table 3). This is in marked contrast to accuracy results obtained when obese subjects actually listened to a tape while correcting proof. While moderately interesting and highly compelling tapes heard during proofreading greatly disrupted the performance of overweight subjects, these identical tapes played before the task had virtually no effect on performance.

Reaction Time

It has been demonstrated that obese subjects correcting proof while distracted are far less accurate than normals. Now let us consider the effects of distraction on another kind of task—reaction time. Median latency to respond to a light stimulus on all errorless trials and the mean number of errors made provide measures of reaction time accuracy. Due to a mechanical failure, undetected until late in the experiment, the reaction time

TABLE 4

Mean of Median Reaction Time Latencies (Msec.)

Subjects	No distraction	Emotional distraction
Normals	462.5	492.9
Obese	403.9	524.3

Comparison

Obese: No Distraction vs. Emotional Distraction $t = 4.49, p < .001$.
No Distraction: Normals vs. Obese $t = 2.19, p < .05$.

data for a large number of subjects were invalidated. It proved impossible to get sufficient additional subjects to make all of the experimental comparisons originally planned.[2] The means for these groups are presented in Table 4, where it is clear that obese subjects were greatly affected by the emotional tape. While their response times when undistracted were shorter than those of normals, they took longer than did normals to respond when the emotional tape was playing (Weight × Distraction $F = 4.53$, $df = 1/36$, $p < .05$). Normals were slightly and nonsignificantly more distracted by the emotional tape when compared to normals without distraction. Obese subjects hearing the emotional tape responded significantly more slowly than did overweight controls performing the task without distraction ($p < .001$).

The number of errors made by subjects in each condition provides a second measure of distraction. An error was recorded when a subject responded with both hands or the incorrect hand. Since only latency scores on errorless trials were included in the first measure, these two indices are independent. The group means are presented in Table 5. While obese and normal subjects made an almost identical number of errors without distraction, obese subjects had a significantly greater number of errors than did normals when hearing the emotional tape.

There is, then, consistency between these data and those provided by the proofreading measures. Again, the obese were strongly distracted by an emotional tape, as reflected in longer response latencies and a greater num-

[2] Rather than totally abandoning the possibility of testing our hypothesis on reaction time, we put all of the new subjects into the Emotional Distraction and No Distraction treatment groups.

TABLE 5
Mean Reaction Time Errors

Subjects	No distraction	Emotional distraction [a]
Normals	7.2	4.7
Obese	7.4	9.1

[a] Comparison of emotional distraction: Obese vs. Normal $t = 2.17, p < .05$.

ber of errors, while normal-weight subjects were not significantly distracted. These findings suggest that the obese are more responsive to competing cues than are normals in at least two different and unrelated test situations.

Emotionality

It is clear from the reaction time and proofreading data that the emotional tapes profoundly affected the performance of the obese. In each case, their accuracy scores were greatly reduced when they listened to the disturbing tapes. In addition to actual performance scores, there were considerable self-report data about physiological and emotional responses to the tapes. These data, presented in Chapter 3, strongly support the assumption that overweight subjects were more affected than were normals by the emotional cues. They reported more arousal than normals when listening to a tape about Hiroshima or dying of leukemia, and they reported substantially less arousal when hearing nonemotional tapes about rain or sea shells. Whether in actual performance or in self-report measures, obese subjects are more affected than are normals by emotionally-arousing stimuli.

DISCUSSION

The present experiment was designed to test implications of the stimulus-binding hypothesis. It was predicted that if overweight individuals are gripped by all prominent external stimuli—not only those relevant to food—they should be distracted by salient competing cues and, in this experimental context, their performance should suffer. The results indicate that the stimulus binding hypothesis is applicable to more than eating behavior. The disruption for the obese, in both reaction time and proofreading, was clear and striking, strongly suggesting that overweight individuals are highly distractible.[3] In addition, cues must be immediate to distract the obese. These results support, for a non-eating behavior, Johnson's (Chapter 8) findings that the presence of visible food cues had a much greater effect on the obese than the manipulation of prior taste in which the food cues, once eaten, were also simply more remote. Similarly in this study, the presence of audible tapes had a disruptive effect on the obese whereas the same stimuli, once completed, became less salient and did not interfere with performance.

If these findings have generalizability outside the laboratory, there should be other indications of distractibility in the obese. For example, the conditions under which they can study effectively should differ. One might expect that the obese would isolate themselves in order to avoid possible distraction. To test this, a group of our subjects was asked to describe the conditions under which they usually studied. Eighty per cent of the obese subjects stated a preference for working in silence and privacy, and only 20% claimed that they worked with a radio playing or other people present. For normals, 58% preferred quiet, and 42% reported that they typically worked with music playing ($\chi^2 = 4.62$, $p < .05$). These results nicely support the major findings of the study by suggesting that overweight subjects' choice of quiet derives from their demonstrably greater distractibility.

Since the distracting input, introduced as part of a secondary thinking task, was relevant to the experiment, it is unclear whether overweight subjects were simply unable to do two things at once or to shut out competing stimuli. Rather than elaborating these two possibilities, neither of which can be conclusively proved or discounted in the present study, let us consider whether there is other evidence that the weight groups differ in the extent to which a cue compels attention and evokes a response. Attention to the primary task is appropriate for this purpose, and performance may be taken as an index of attention to the cues relevant to monitoring lights and correcting proof. As shown, performance in both tasks was considerably faster for overweight than normal subjects when undistracted. Rodin et al. (Chapter 12) found similar results for other attentional measures. In addition to performance scores, there are converging data in the questionnaire responses which also suggest that the obese were more attentive to the tasks than normals.

[3] We are of course interpreting these results as supporting the hypothesis that the interaction between responsiveness and stimulus prominence in the obese applies to areas other than consummatory behavior. Although somewhat far fetched, the greater disruptive effect of salient distractors on obese subjects might also be explained by differential motivation between weight groups if it is assumed (a) that the more absorbing the distractor, the greater would be a subject's perception that the experimenter wished him to attend to the distractor and (b) that obese subjects complied more than did normals. At least one of the experiments described by Pliner in Chapter 14 (time perception) was originally designed (Pliner, 1973a) to test the interaction between cue salience and responsiveness in an experimental context in which this alternative explanation is even less plausible. This study supports the cue prominence-reactivity hypothesis as opposed to an explanation based solely on compliance differences.

Subjects were asked, after each task, to write down their thoughts while proofreading or tracking the lights. It was assumed that thinking about the task could be taken as an index of attention. To consider this possibility, reported thoughts were divided into four categories.[4] (*a*) those directly related to the reaction time or proofreading, (*b*) those related to the experiment in general (the purpose, the test room, the tapes, etc.), (*c*) thoughts about events which were not immediate or relevant to the experiment, e.g., future papers or exams, girlfriends, etc., and (*d*) no response.

In all conditions where subjects worked without distraction (No Distraction, Pre-task Neutral, and Pre-task Emotional), there was a significant correlation between thinking exclusively about the task and a high level of performance (point biserial $r = .4319$, $p < .01$). While it is clear that those subjects who attended to, or at least thought about, the tasks did better, the important question is whether there were more overweight than normal subjects in this group. Twenty-two obese subjects, and only 14 normals, reported task-related thoughts. In addition, not only were the obese overrepresented in the sample of subjects thinking exclusively about the task, they were also underrepresented in the group of subjects reporting exclusively stimulus-independent thoughts, i.e., 11 normals and only 2 obese subjects reported thoughts for which external cues were not immediately present. The 2×2 chi square for task vs. self-reported thoughts equals 6.27, $p < .02$.

The data suggest that when a cue is immediate and compelling, the obese attend to it and think about it; however, they actually appear to think less than normals about events which are not immediate. These results support the findings for the effects of Pre-task Distraction which also indicate that obese subjects are primarily attentive to current stimuli. In these conditions where the tapes preceded the tasks, the taped materials had virtually no effect on the obese subjects, and the task itself appears to have gripped their attention.

Finally, the results show that, like VMH-lesioned animals, overweight people are highly reactive to compelling emotional stimuli. For emotionality, like eating, thinking and attention, the obese appear stimulus bound. It is tempting to consider the engaging predictions which can be derived, with not too much fancy footwork, from these findings. On the whole, obese individuals whose thinking seems to be stimulus bound should be less creative than normals. On the other hand, they should be marvelous accountants, at least when undistracted. Given a salient exciting stimulus, the obese should be more likely than normals to become sexually aroused. While these speculations could continue, the point, by now, is clear. The present experiment has demonstrated that overeating is just one aspect of a fat person's hyperresponsiveness to a wide variety of potent external stimuli. A good portion of his behavior may be predictable from this proposition.

[4] Responses were scored by two coders independently. Ratings agreed in 93% of the cases. In cases of disagreement, the ratings which produced the most conservative appraisal of the hypothesis were always used.

14
ON THE GENERALIZABILITY OF THE EXTERNALITY HYPOTHESIS[1]

Patricia Pliner

The results of the experiments by Rodin (Chapter 13) and by Rodin, Herman, and Schachter (Chapter 12) are consistent with the hypothesis that the responsiveness exhibited by the obese to external food cues is only one aspect of a more general responsiveness to external cues. Moreover, Rodin's data also appear to confirm, in a noneating context, Ross' corollary (Chapter 7) that the salience of the external stimulus is an important parameter. Only when the external cues were highly salient did the obese respond more strongly than normals; with cues low in salience, the obese were no more responsive and, indeed, appeared to be slightly less responsive than normals.

These first indications that the aberrant eating behavior of the obese may be a special case of a much broader phenomenon are particularly intriguing for, if correct, there are major implications for our conception of the nature and origin of obesity. It would seem of importance, then, to investigate the general externality notion on as wide a variety of nonfood-related behaviors as possible. The present paper will present the results of three experiments which further test the generalizability of the external sensitivity ideas in a variety of experimental contexts having nothing to do with consummatory behavior.

THE EFFECTS OF AUDITORY CUES ON TIME-ESTIMATION JUDGMENTS

In the first experiment, Pliner (1973a) presented subjects with auditory cues and manipulated their salience by varying loudness. The criterion

[1]The research reported in this chapter was supported by Canada Council Grant No. S70-1571. The author wishes to acknowledge Patricia Meyer and Mary Anne Hadden for the exceedingly competent experimental assistance and Howard Cappell for his valuable comments on an earlier draft of this manuscript.

response was time estimation; the rationale for its use is derived from recent work of Ornstein (1969). Ornstein hypothesized that the subjective experience of the duration of an interval of time is a direct function of the amount of information stored in memory during that interval. Thus, any procedure which alters the "size of storage" should also affect the experience of the duration of that interval. For example, increasing or decreasing the amount of input by manipulating the number of external stimuli present in a given interval should affect duration experience. To test this derivation Ornstein presented subjects with three 9.5 min. tape recordings. The tapes contained 373, 746, or 1,119 auditory stimuli consisting of .2 sec. tone pulses. After a tape had been played, the subject was asked to estimate its length. The data showed a strong positive relationship between the number of stimuli on a tape and subjects' estimates of elapsed time. The more stimuli presented during an interval, the longer it was judged.

For Ornstein, then, time experience is very closely related "to the mechanisms of attention, coding, and storage in the brain," and any experimental manipulations which affect these processes should affect time experience. It is individual differences in precisely these processes on which Rodin, Herman, and Schachter (Chapter 12) focused in conceptualizing the notion of differences in sensitivity to salient external cues. According to Rodin et al., "it may be that . . . the obese . . . actually take in more information than normals due either to sensory or attentional differences, encode it more effectively, or store more in short-term memory." Thus, given that the mechanisms of attention, coding, and storage of external stimuli are involved in time experience, and given that obese and normal individuals may differ in terms of these processes, it might be expected that there will be obese-normal differences in time experience. More specifically, if the obese are more responsive than normals to salient external stimuli, they should store more information in memory than normals during exposure to such stimuli, and if they are equally or less responsive than normals to stimuli low in salience, they should store a comparable or lesser amount in memory than normals during exposure to such stimuli. Since it has been shown that time experience is a function of amount of information stored in memory (Ornstein, 1969), these derivations would lead to the following specific predictions:

1. When presented with a series of highly salient external stimuli, obese subjects will estimate the duration of elapsed time as longer than normals.
2. When presented with a series of external cues low in salience, obese subjects will estimate the duration of elapsed time as no longer and perhaps shorter than normals.

Overview of Experimental Procedure

On the pretext of precisely timing their exposure to a visual stimulus, obese and normal subjects were exposed to tape recordings of auditory stimuli which varied in duration (4 minutes and 8 minutes), number of stimuli (40/minute and 80/minute), and salience (45 db. or 90 db.). Next, subjects were presented with a questionnaire which contained, among others, items asking that they estimate the length of the stimulus tapes and that they rate them on 7-point scales in terms of loudness and annoyance.

The four independent variables were manipulated orthogonally in a $2 \times 2 \times 2 \times 2$ factorial design.[2] There were to have been five subjects in each of the 16 experimental conditions;[3] however, a limited supply of subjects precluded running any more than the 70 on whose data this report is based.

Stimuli and Presentation of Stimuli

The auditory stimuli consisted of tape recordings of .2 sec. tone pulses of 500 c.p.s. frequency, produced at a rate of either 40 or 80 pulses per minute. For each rate of stimulus presentation, a four-minute and an eight-minute tape was prepared.

During the experiment, the stimulus tapes were played on a tape recorder placed directly behind the subject at a distance of 34 inches. The salience of the auditory stimuli was controlled by adjusting the volume dial of the tape recorder to one of two predetermined locations, which produced sounds measured at 45 or 90 db. Thus, salience is manipulated in terms of the physical properties of the stimulus. Just as Ross (Chapter 7) assumed that the salience of a visual cue varied with the brightness of illumination, it is assumed here that the salience of an auditory cue varies with loudness. During presentation of the stimuli, subjects were seated in a dentist's chair located inside a sound-attenuation chamber.

Procedure

The study was presented to the subjects as an investigation of the relationship between cognitive and physiological reactions to visual stimuli. They were informed that their physiological responses would be measured by means of heart-rate electrodes attached to their wrists, while their cognitive responses would be measured by questionnaires. After the electrodes were attached, each subject was told that it was crucial that all subjects be exposed to the visual stimulus for exactly the same length of time; for that reason,

[2]Initially, it was intended that weight and cue potency be treated as between-subject variables while tape duration and number of stimuli act as within-subject variables. Each subject would thus have been run in each of the four possible variations of tape duration and number of stimuli at one of the levels of cue potency. Unfortunately, a pretest of this procedure and a discussion with subjects indicated that subjects were attempting to be consistent and for that reason were giving identical time estimates for each of the four conditions. It is for this reason that the design was changed to an entirely between-subject design.

[3]A sixth normal-weight subject was inadvertently run in one of the conditions (4 minutes, 80 stimuli/minute, 90 db.). His data have been included in all analyses; however, the results of the study are virtually unchanged if this subject's data are not included.

before being presented with the visual stimulus, he was to close his eyes and open them only "when some tape-recorded beeps begin." He was to look at the visual stimulus until the "beeps" stopped and was then to close his eyes again. In other words, the purported purpose of the "beeps" was to time the subject's exposure to the visual stimulus. These "beeps" were, of course, the auditory stimuli comprising the true manipulatiion of the study. Following presentation of the irrelevant visual stimulus (TAT card #8BM) and the accompanying auditory stimuli, subjects were given a questionnaire entitled "Cognitive Reaction Questionnaire" which contained, in keeping with the cover story, a large number of questions about the visual stimulus. Embedded among them were the following items:

How long was the period during which the tape (beeps) was on?

Circle the number of minutes.

1 2 3 4 5 6 7 8 9 10 11 12 13 14 15 16

How loud did you find the beeps?

Extremely Loud ____ ____ ____ ____ ____ ____ ____ Not at all Loud

How annoying did you find the beeps?

Extremely Annoying ____ ____ ____ ____ ____ ____ ____ Not at all Annoying

Subjects. Subjects were 70 males recruited through newspaper advertising and questionnaires in several high schools in suburban Toronto. A subject was considered to be obese if his actual weight was 15% or more above the average weight for his height as indicated by a standard table of Canadian heights and weights. A subject was classified as normal if his actual weight was no more than 10% above the average weight for his height. Overweight subjects averaged +32.1% overweight, and the normal group had a mean weight of −3.5% overweight.

Results

Check on the loudness manipulation. Analysis of variance of responses to the question "How loud did you find the 'beeps' " showed that subjects rated the high-salience stimuli as significantly louder than the low salience stimuli ($\bar{F} = 22.27$, $df = 1/55$, $p < .01$). Thus, there is no doubt that the salience manipulation was successful. Moreover, there were no significant main effects or interaction effects involving weight.

Rated annoyance of stimuli. The "annoyance" question was included to check for the possibility that any observed differences in time estimations could be accounted for by differences in some emotional effect of the stimuli rather than by differences in salience. If subjects found the 90 db. tones more annoying than the 45 db. tones, time-estimation results in accordance with

TABLE 1
Mean Time Estimation in Minutes [a]

Time	Obese			Normal		
	High salience	Low salience	High & low salience Combined	High salience	Low salience	High & low salience Combined
4 Minute	3.50 (8)	2.38 (8)	2.94 (16)	3.45 (11)	4.56 (9)	3.95 (20)
8 Minute	7.22 (9)	4.88 (8)	6.11 (17)	4.67 (9)	5.62 (8)	5.11 (17)
4 and 8 Minute Combined	5.47 (17)	3.62 (16)	—	4.00 (20)	5.06 (17)	—

[a] The number in parentheses refers to the number of subjects in that condition.

predictions could be interpreted in terms of emotionality as well as in terms of salience. Analysis of variance of the "annoyance" data showed only two significant effects: The eight-minute tapes were rated as more annoying than the four-minute tapes ($F = 7.35$, $df = 1/55$, $p < .01$), and there was a significant interaction between the rate of stimulus presentation and weight ($F = 6.05$, $df = 1/55$, $p < .05$). Correlations between the annoyance ratings and time estimations were: obese subjects, $r = -.17$; normal subjects, $r = +.06$. Neither correlation approached statistical significance. Thus, it appears unlikely that time-estimation results in accordance with predictions could be alternatively interpreted in terms of emotionality rather than salience.

Time Estimations. Before considering the effects of obesity and salience, let us examine whether or not Ornstein's phenomenon is a reliable one. It is. Subjects who heard tapes containing 80 stimuli/minute estimated that they lasted 5.19 minutes. Subjects who listened to the 40 stimuli/minute tape estimated that they lasted 3.79 minutes. This difference is significant with $F = 18.48$, $df = 1/55$, $p < .001$. It is the case that time estimation is strongly affected by the sheer number of external stimuli present in a given interval.

Given this fact, let us examine the effects of obesity and cue salience on time estimation. The basic data are presented in Table 1. Fig. 1 plots the effects of salience and obesity on time estimation. It is evident that there is a highly significant ($F = 8.65$, $df = 1/55$, $p < .01$) interaction between weight and cue salience.

As predicted, in the high-salience condition, obese subjects estimated the time elapsed as significantly longer than did normals ($t = 2.17$, $df = 55$, $p < .05$), while in the low salience condition the direction of the difference was reversed ($t = 2.03$, $df = 55$, $p < .06$). Thus, the main predictions of the study were strongly confirmed.

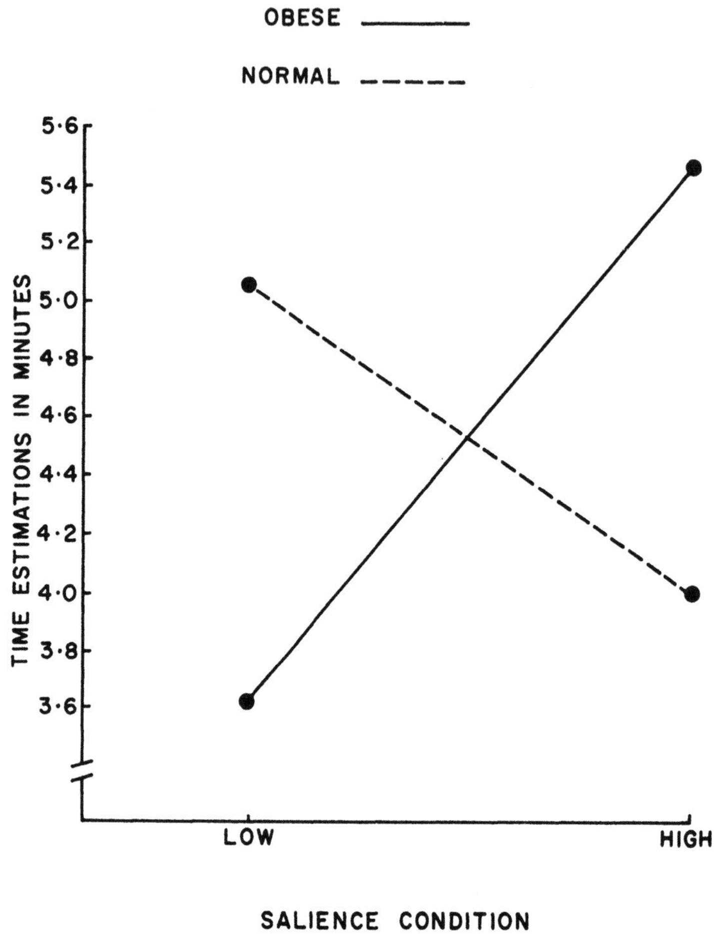

Fig. 1. Time estimation as a function of cue salience.

A final and unexpected effect was the significant interaction ($F = 4.12$, $df = 1/55$, $p < .05$) between tape duration and weight, which is depicted in Fig. 2. The configuration of this interaction is similar to that presented in Fig. 1, with obese subjects estimating the eight-minute tapes as longer and the four-minute tapes as shorter than normals.

Discussion

The results of the first experiment strongly confirmed the prediction that cue salience and weight would interact in determining responsiveness to an auditory external cue. With cue salience varied by manipulating loudness

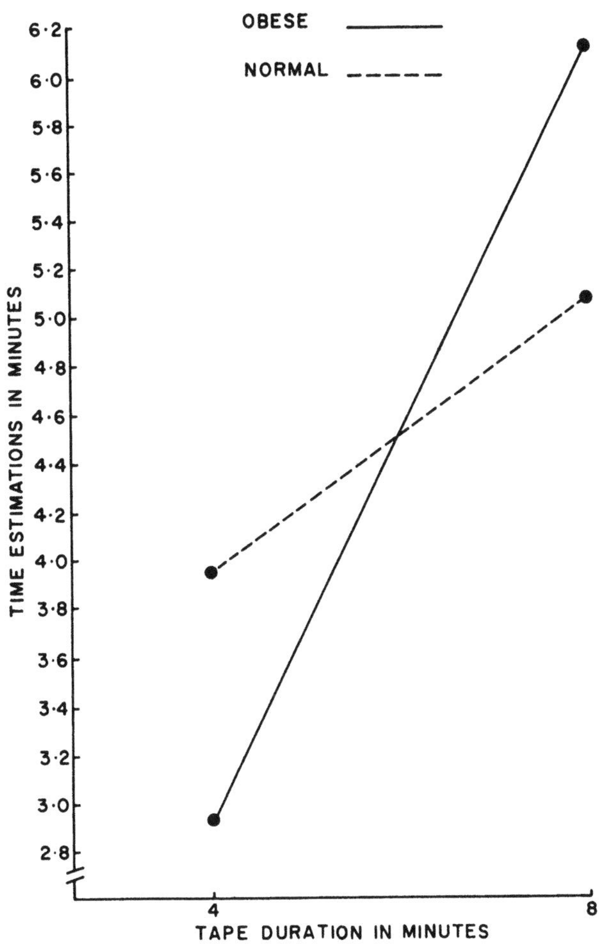

Fig. 2. Time estimation as a function of stimulus tape duration.

and with time estimation as the response, obese subjects were more responsive than normals to highly salient external cues but less responsive than normals to cues low in salience. Since neither of the significant effects of the ratings of the "annoyance" of the stimuli coincided with this interaction in time estimations, it appears unlikely that the results could be interpreted in terms of some emotional aspect of the stimuli rather than their salience. While the finding of a similar interaction between tape duration and weight was not predicted, it can be similarly explained with a simple assumption. That is, an eight-minute collection of stimuli may be more salient or prominent than a four-minute collection; accordingly, obese subjects responded more strongly than normals in the former case and less strongly in the latter.

THE EFFECTS OF VISUAL CUES ON THINKING BEHAVIOR

In a second experiment, Pliner (1973b) manipulated the salience of a visual scene and then measured the extent to which subjects actually thought about the scene. All subjects were asked to think about a particular scene; some subjects were then provided with salient external cues in the form of a slide depicting this scene, while in a low-salience condition subjects were merely provided with a description of this scene. A third group of subjects was not asked to think about the scene at all. It was predicted that:

1. When the scene was highly salient, obese subjects would report themselves to have spent more time thinking about it than normals.
2. When the scene was low in salience, obese subjects would report themselves to have spent no more time and perhaps less time thinking about it than normals.

In addition to this direct measure of thinking behavior, an indirect measure was also used. During the thinking-periods subjects were exposed to a potentially painful ice-water stimulus. Under the assumption that thinking about the scene would constitute a distraction from the cold-produced pain (see August, 1961; Greene & Rehner, 1972; Roazin, 1967; Szasz, 1957), it was further predicted that:

3. When the scene was highly salient, obese subjects would be more distracted and would, therefore, report pain from the ice water later than normals.
4. When the scene was low in salience, obese subjects would be no more distracted and would, therefore, report pain from the ice-water no later and perhaps sooner than normals.

Overview of Experiment

Under the pretext of measuring their physiological responsiveness to cold, obese and normal subjects were required to immerse their hands in an ice-water solution maintained at 0-4° C during 6 three-minute trials. Crosscutting the weight variable were three experimental conditions:

1. High Salience—during ice-water trials subjects were asked to think about a scene which was flashed on a screen during every trial.
2. Low Salience—during ice-water trials subjects were asked to think about a scene which was described to them before each trial.
3. No Salience—during ice-water trials subjects were not asked to think about a particular scene.

The dependent variables included self-reports of how long subjects thought about the scene and a measurement of latencies of report of cold-produced pain.

Procedure

Upon his arrival at the laboratory, each subject was given an expanded version of the following cover story:

> The purpose of today's experiment is to measure the physiological effects of cold on human subjects. Cold is considered by psychologists to be an important source of stress—important because nearly everyone is exposed to cold at one time or another. I'll be attaching leads to the lower part of your arm so we can monitor your heart rate continually during the study. The way we are going to expose you to cold is very simple. I'll just have you dip your hand into a bucket of cold water. At the same time I'll be measuring your heart rate, and that, basically, is all there is to the experiment.

At this point electrodes leading to an impressive-looking but totally nonfunctioning piece of apparatus were attached to the wrist of the subject's dominant hand and the instructions continued:

> . . . In this experiment as in all experiments it is important that everything be as similar as possible from one subject to the next. One important potentially uncontrolled source of variation is what subjects are thinking about. What you are thinking about can have a large effect on your heart rate. For instance, if you were thinking about an upcoming exam, that might affect your heart rate. Now I cannot control your thoughts but what I am going to do is to give you something to think about. There will be six cold-water trials in all and during each trial . . .

At this point subjects in the different conditions received different instructions

> *High Salience* . . . I am going to turn out the lights and turn on this projector and give you a scene to think about. I want you to concentrate as much as you possibly can on the scene and not to think about anything else. Your mind may wander but try to think about the scene as much as you can.
>
> *Low Salience* . . . I am going to turn out the lights and turn on the projector motor to mask noises outside the lab and read a descriptive paragraph to you. I want you to try to imagine the scene in your mind's eye. Your mind may wander but try to think about the scene as much as you can.
>
> *No Salience* . . . I am going to turn out the lights and turn on the projector motor to mask noises from outside the lab. I am going to ask you to try not to get involved with any noises you may hear from down the hall. Your mind may wander but try to keep your thoughts on what is happening here as much as you can.

Each trial began with the subject immersing his nondominant hand in water maintained at approximately 35° C for two minutes. Following this preparatory period he listened to the instructions for his condition and then put his hand into water maintained at 0–4°C for three minutes.

On each trial immediately before immersing one hand in the ice water, the subject was instructed, "when you first experience what you would call pain, say 'now.'" Then he filled out the questionnaire described below and replaced his hand in the warm water to begin the next trial.

Scenes. After extensive pretesting, two scenes which met the following requirements were chosen to be used as external stimuli:(*a*) high mean ratings on bipolar scales on dimensions of pleasantness, interest, likeability, and "relaxingness" and low variability on these dimensions; (*b*) similarity of ratings obtained from obese and normal individuals; and (*c*) elicitation of similar descriptions from individuals asked to describe them.

For each of the two scenes, a single description was prepared based on a composite of the pretest subjects' descriptions. An example follows:

> I want you to imagine an ocean at sunset. There is a strip of beach in the foreground. On the right hand side of the beach, a man is silhouetted against the sea and sky. A small wave is about to break on the beach, and beyond the wave two large rocky islands rise out of the quiet sea. The sky contains a line of clouds which accentuate the colors of the setting sun. The whole scene has a reddish hue; the long line of clouds is dark orange, the sky is orange and red, and the sea reflects those colors. The rocks and the human figure provide a sharp, black contrast.

The other scene was of a mountain locale and its description was similar to that of the first both in terms of the number of words and the amount of imagery. Each of the two scenes served as an external cue for half of the subjects in the High and Low Salience conditions.

Measurement. There were two main dependent variables in the study. First, at the end of each trial subjects answered the following question: "What percent of the time that your hand was in the water did you spend thinking about the (mountain or beach) scene?"[4] Second, subjects were instructed before each trial: "When you first experience what you would call pain, say 'now'." The length of time elapsing between immersion in the ice water and this point (now) was measured with a stopwatch. In addition, at the end of each three-minute ice-water trial subjects filled out a questionnaire containing the following items:

1. What percentage of the time that your hand was in the ice water did you spend thinking about the experimental situation (the cold, the pain, etc.)?
2. What percentage of the time that your hand was in the ice water did you spend thinking about other things?
3. How painful did you find the cold water stimulus?

[4]All questions involving mention of the (mountain or beach) scene were omitted for subjects in the No Salience condition since they received no instructions to think about a scene.

4. How effortful did you find thinking about the scene?
5. How interesting did you find thinking about the scene?

Subjects. Subjects were 72 undergraduate male students who were recruited from psychology classes, and they were paid $1.50 for their participation. The same criteria as in the time-estimation study were used for classifying subjects into weight groups. Obese subjects ranged between +15.0% and +71.1% overweight and normals between −12.5% and +9.8%.

Results

To review the expected results briefly: If it is the case that the obese are more responsive than normals to salient external cues, they should report themselves to have thought about the visual scene more than normal subjects in the High Salience condition but no more and perhaps less than normals in the Low Salience condition. It was further predicted that obese subjects would report pain later than normals in the High Salience condition but no later and perhaps sooner in the Low Salience condition. The No Salience control condition was included merely to ensure that obese and normal subjects did not differ in the latency with which they reported pain in the absence of any instructions to think about a particular scene.

A note on the analysis of variance designs. In the analyses of variance to follow, slightly different designs were used, depending on the particular dependent variable under consideration. The data on time spent thinking about the scene were analyzed using a 2 × 2 × 2 × 6 repeated measures design. The factors were weight (Obese vs. Normal), Salience condition (High Salience vs. Low Salience), scene (Mountain vs. Beach), and the six trials. The No Salience control condition was not included in this analysis since subjects in that condition were not asked to think about a particular scene.

For the remaining analyses of variance a 2 × 3 × 6 repeated measures design was used. The factors were weight (Obese vs. Normal), salience condition (High vs. Low vs. No Salience), and the six trials. Scene (Mountain vs. Beach) was not taken out as a factor since it was irrelevant in the No Salience condition.

Self-report of time spent thinking about the scene. The analysis of the responses to the question "What percentage of the time that your hand was in the water did you spend thinking about the (mountain or beach) scene?" yielded two significant effects. First, there was a significant effect of trials ($F = 5.05$, $df = 5/200$, $p < .01$); subjects spent progressively less and less time thinking about the scene on successive trials. On the first trial they reported themselves to have spent on the average 52.8% of the three-minute trial thinking about it; this had decreased to 38.1% by the last trial.

Second, and in the present context more interesting, was the significant interaction between weight and salience condition ($F = 4.08$, $df = 1/40$, $p < .05$).

The relevant means are depicted in Fig. 3 where it can be seen that in the High Salience condition obese subjects spent a greater proportion of time thinking about the scene than did normals ($\overline{X}_{Obese} = 53.2\%, \overline{X}_{Normal} = 42.8\%$), while in the Low Salience condition, this difference was reversed $\overline{X}_{Obese} = 35.1\%, \overline{X}_{Normal} = 47.4\%$).

Pain latencies. Just as in the thinking-time data, there was a significant interaction between weight and salience condition in the latency with which pain was reported ($F = 4.59$, $df = 2/66$, $p < .05$). It can be seen in Fig. 4 that in the High Salience condition obese subjects had on the average longer pain latencies than normal subjects ($\overline{X}_{Obese} = 80.9$ sec., $\overline{X}_{Normal} = 56.0$ sec.), while in the Low Salience condition these differences were reversed ($\overline{X}_{Obese} = 40.5$ sec., $\overline{X}_{Normal} = 89.2$ sec.). In the No Salience control condition, pain latencies for obese and normal subjects were quite similar ($\overline{X}_{Obese} = 59.0$, $\overline{X}_{Normal} = 64.2$ sec.).

Also significant was the effect of trials ($F = 16.64$, $df = 5/330$, $p < .01$); pain latencies became longer with each successive trial ($\overline{X}_{trial\ 1} = 40.7$ sec.; $\overline{X}_{trial\ 6} = 83.8$ sec.).

Topics of thought. After each trial subjects in all conditions were asked to indicate the proportion of the time they spent thinking about things other than the experiment itself and (in the High and Low Salience conditions) the (mountain or beach) scene. Although the mean proportion of time spent thinking about other things was only 20.0% of the total,[5] 32 normal and 35 obese subjects had, in fact, thought about other things. These subjects were asked to indicate briefly what they had been thinking about and their responses were coded into three categories by a coder, who was blind to the subjects' weights and to the experimental conditions:

1. Externally-generated thoughts—thoughts about objects which were physically present and visible in the experimental situation and therefore presumably high in salience, e.g., the floor tiles, the wrinkled skin caused by the water.

2. Self-generated thoughts—thoughts about objects and events which were not present and visible in the experimental situation and therefore low in salience, e.g., girlfriend, summer job.

3. Mixed thoughts—thoughts which could not be classified as externally-generated or self-generated. They were not directly about objects in the experimental situation but appeared to be derived from or prompted by the experimental situation, e.g., making snowballs with no gloves on, a science fiction story about cold stress.

Responses in these categories were summed across trials for each of the 67 subjects, resulting in three scores for each subject. Only 4 subjects reported exclusively externally-generated or self-generated thoughts; even when the

[5]Obese and Normal subjects did not differ on this measure, nor were there any interactions involving weight.

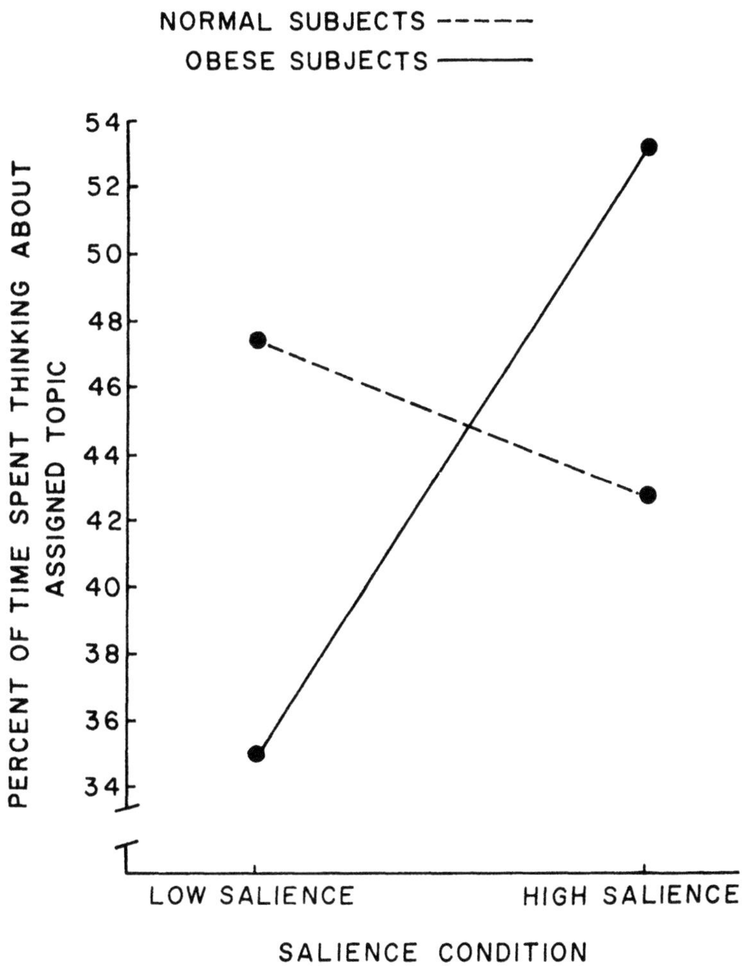

Fig. 3. Self-report of time spent thinking about the (mountain or beach) scene as a function of salience condition.

mixed category was ignored, fewer than half of the subjects entertained only one kind of thought. Subjects were, therefore, classified into those for whom the number of externally-generated thoughts was greater than or equal to the number of self-generated thoughts and those for whom the number of externally-generated thoughts was less than the number of self-generated thoughts. The data are presented in Table 2; analysis of the data yielded a χ^2 of 5.33 ($p < .05$). Thus obese *S*s were more likely to think about highly salient objects and events than those low in salience, while for normal *S*s the direction of this difference was reversed.

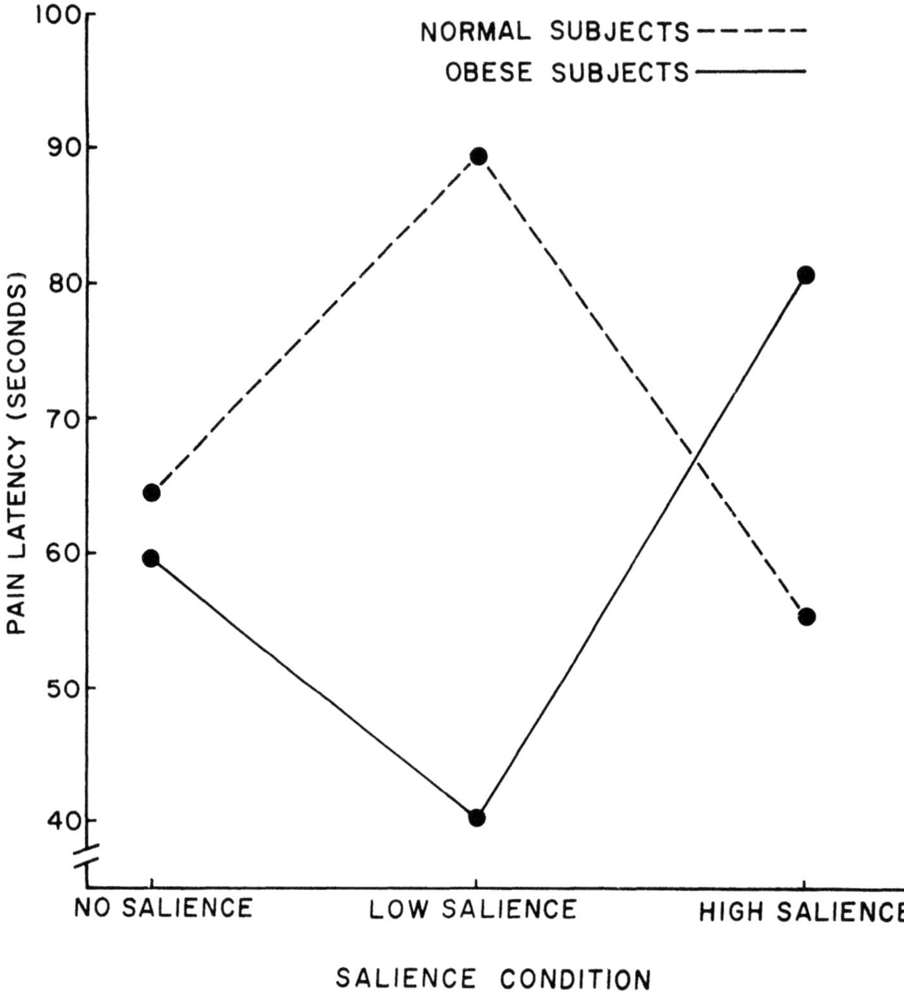

Fig. 4. Pain latency as a function of salience condition.

Pain ratings. Analysis of the ratings of the "painfulness" of the ice water obtained at the end of each trial showed only one significant effect; the water was rated less painful on each successive trial ($F = 12.52$, $df = 5/330$, $p < .01$). The mean for Trial 1 was 3.41[6]; the mean for Trial 6 was 5.91. No other effect even approached statistical significance.

[6]The ratings were obtained by having the subjects mark an undifferentiated 14 cm. line labeled at the endpoints "extremely painful" and "not at all painful." The data are reported in cm. from the "extremely painful" end of the scale.

TABLE 2
Relative Frequency of Externally-Generated and Self-Generated Thoughts for Obese and Normal Subjects

	Number of subjects for whom:	
	Externally-generated $\geqslant$ self-generated	Externally-generated $<$ self-generated
Obese	23	12
Normal	12	20

Discussion

The data of the second experiment strongly supported the prediction that the thinking behavior of the obese would be more responsive to salient external cues than that of normals. Whether thinking behavior was measured directly by self-report or indirectly by distraction from pain, obese subjects thought more about a highly salient visual stimulus than normals and thought less about a stimulus low in salience than did normals. In addition, to these data which showed the effects of manipulated external cues on behavior, analysis of subjects' reported thoughts when they were thinking about "other things" showed a correlation between weight and responsiveness to salient external cues. These data replicate Rodin's (Chapter 13) finding of the greater effects of external cues on the reported thoughts of overweight subjects than of normals.

There were, however, some data in the study which at first glance are puzzling. It was, of course, assumed in designing the study that the self-report of thinking-time and the pain-latency data were both manifestations of the same process—the extent to which subjects were thinking about the scene. Given this assumption, it was somewhat suprising to discover that the trials effects for these two dependent variables appeared to be in opposite directions. Subjects reported themselves to be thinking less about the scenes over trials yet showed greater pain latencies over trials. Since these two variables are presumably determined by the same thing, one would expect them to be positively related or perhaps unrelated, but certainly not negatively related. That this paradox is only apparent was demonstrated by examining separately for each trial the correlation between self-report of thinking time and pain latency. The correlations for Trials 1 to 6 are respectively: +.29, +.11, +.33, +.15, +.35, and +.22; the mean correlation is +.25 ($df = 46$, $p < .05$). Thus, while from the trials effects in the analyses of variance the self-report of thinking-time data and the pain-latency data appear to be negatively related, they are actually positively related within any one trial.

Comparisons of pain ratings for obese and normal subjects in the No Salience control condition made it possible to determine whether obese and normal subjects differed generally in responsiveness to painful stimuli. Any such differences would have seriously undermined the pain-latency data; however, no such differences occurred.

It might have been expected that the ratings of the "painfulness" of the ice-water stimulus obtained at the end of each trial should have been affected by the experimental manipulations in the same way as were the pain latencies. However, neither the salience condition nor the interaction of salience condition with weight had any measurable effect on the ratings. This lack of effect becomes somewhat more understandable if the cold pressor situation is examined more carefully. Data from Barber and Hahn (1962) indicate that pain in a cold pressor situation increases as immersion time increases, becoming quite severe during the third minute. Since it has been shown that, at least in one situation, cognitive manipulations have little effect on severe as opposed to mild pain (Nisbett & Schachter, 1966), it is not surprising that the cognitive manipulations of the present study had little effect on ratings taken immediately after the point of most severe pain. The pain latencies, obtained before the pain was severe, were affected by the manipulations, but the pain ratings obtained later were not.

POSITIVE AND NEGATIVE EMOTIONALITY

The third experiment investigated the heightened responsiveness of the obese with respect to emotionality. Studies employing self-ratings of emotionality have shown that obese subjects are more upset than normals by such aversive stimuli as the threat of electric shock or by tape-recorded accounts of the bombing of Hiroshima and death from leukemia (Chapter 3). Using a behavioral measure of emotional responsiveness, Rodin (Chapter 13) found that obese subjects spent a greater proportion of time working to avoid electric shock than did normals. Rodin, Elman, and Schachter (Chapter 3) also found that the performance of obese subjects in a maze learning task was more disrupted by shock than was that of normals.

It should be noted that this generalization about responsiveness to emotional stimuli is based entirely on studies using aversive stimuli, i.e., threat of electric shock, actual electric shock, and tape recordings relating upsetting material. Although the evidence is by no means unequivocal, psychological and psychiatric assessments often show that the obese score higher on measures of anxiety and depression than do individuals of normal weight (e.g., Moore, Stunkard, & Srole, 1962). If this is the case, then hyperresponsiveness to aversive emotional stimuli may simply be a manifestation of this greater anxiety and depression and not a manifestation of a heightened responsiveness to affective cues in general. If it can be shown that the obese are more responsive than normals to *positive* affective stimuli as well as negative affective stimuli, this would be evidence that heightened responsiveness to emotional cues by the obese is indeed another aspect of a general heightened responsiveness to external cues.

Method

Subjects. Subjects were 46 male high school and college students who were recruited through newspaper advertising and canvassing in the local schools. Identical criteria to those used in the first two experiments were employed to divide subjects into overweight and normal weight groups. The mean percentage overweight for obese subjects was 30.4% (range = 15.2% to 58.8%) while normal subjects were on the average 0.7% underweight (range = 9.5% underweight to 9.5% overweight).

Procedure. Subjects were informed when they arrived at the laboratory that the purpose of the experiment was to measure their physiological responses to various stimuli. Electrodes appropriate for measuring heart rate, galvanic skin response, respiration rate, and eyeblinks were attached to the subject, and he was then seated in a comfortable reclining chair in a sound-attenuated chamber. After a few minutes during which the subject was presented with a series of auditory stimuli and then rested for five minutes, the present experiment began.

Each subject looked at a series of slides and after viewing each slide filled out seven rating scales.

Rating Scales. The rating scales were standard 7-point scales labeled at the ends with the following pairs of adjectives: disliked-liked, tensing-relaxing, ugly-beautiful, unpleasant-pleasant, unappetizing-appetizing, frightening-calming, nauseating-not nauseating.

The adjectives were chosen to tap an emotional response and at the same time be appropriate to rate all of the stimuli presented.

Stimuli and Presentation of Stimuli. The crucial stimuli to which subjects were exposed were a positive emotional slide, a negative emotional slide, and a neutral slide. In order to conceal the purpose of the experiment, they were embedded in a series including several irrelevant slides. The neutral slide depicted a glacier on a mountain, the positive slide displayed an extremely attractive nude female, and the negative slide showed several bloody human organs on an autopsy table.

During presentation of the slides, the subject was seated in the darkened, sound-attenuated chamber facing a screen directly in front of him. Each slide was presented for 30 seconds, and there was an interval of approximately 60 seconds between slides during which the subject made his ratings. For all subjects the neutral slide was presented first; the emotional slides were alternated so that approximately half the subjects in each weight group saw the negative emotional slide next while the remaining subjects saw the positive emotional slide next.

Results

To review the expected results briefly: If it is true that the obese respond more strongly than normals to emotional stimuli, they should rate the positive affective stimulus more positively and the negative affective stimulus more negatively than subjects of normal weight. There should, however, be no differences between the weight groups in ratings of the neutral stimulus.

In order to evaluate the data, the rating scale scores were summed across the seven adjective pairs for each subject for each of the slides, resulting in a single emotionality score for each subject for each of the three stimuli. The means of these emotionality scores appear in Table 3. The scores were then subjected to 2×3 repeated measures analysis of variance. The interaction between weight and emotional content of the stimulus was highly significant ($F = 5.44$, $df = 2/88$, $p < .025$), showing that subjects in the two weight groups responded differently to the various stimuli. Orthogonal comparisons between appropriate pairs of means showed that obese subjects rated the positive slides more positively ($t = 2.15$, $df = 44$, $p < .05$) and the negative slide more negatively ($t = 1.96$, $df = 44$, $p < .06$) than subjects in the normal group. The difference between obese and normal subjects for the neutral slide did not approach statistical significance ($t = 1.23$, $df = 44$, n.s.)—a finding which indicates that the obese are more extreme in their reactions only to arousing slides.[7]

Discussion

Obese subjects rated a positive emotional stimulus more positively and a negative emotional stimulus more negatively than normals, clearly confirming the hypothesis that the obese respond more strongly than normals to affective stimuli, both negative and positive. Although these data extend the range of stimuli to which the obese are emotionally overresponsive, the generality of the results may be limited by the fact that only one stimulus of each type was used. Nonetheless, the study demonstrates that at least for the

[7] It is, of course, possible that there could be no mean difference between the two groups of subjects in their ratings of the neutral slide and still the obese would be more extreme in their reactions. Such would be the case if the obese subjects as a group were more variable in their reactions to the slide with some being highly favorable and others unfavorable. This could also be the case if the obese subjects individually were more variable in their responses to the 7 scales on which the slides were each rated. The evidence on variability, however, indicates that neither of these alternatives is correct. The group variance for the obese subjects is 6.50 and for normal subjects 8.96. If anything the obese as a group are less variable than normals. Similarly, the average individual variability across scales is smaller for the obese ($\sigma^2 = 1.08$) than for normal subjects ($\sigma^2 = 1.37$).

TABLE 3
Mean Affective Responses of Obese and Normal Subjects to Positive, Negative, and Neutral Affective Stimuli[a]

Subjects	Positive	Negative	Neutral
Obese	11.17	39.65	22.00
Normal	15.48	35.74	24.48

[a]The most positive score possible would be 7; the most negative score possible would be 49.

specific slides employed, overweight subjects are, as predicted by the "general externality" hypothesis, more responsive than are normals.

IS THE EXTERNALITY HYPOTHESIS GENERALIZABLE?

In three separate studies, the relative responsiveness of obese and normal subjects to salient external stimuli was assessed. And in each case, whether the stimulus was a series of auditory tones, a visual scene, or positive and negative affective stimuli, obese subjects responded more strongly than normals. This evidence, taken in combination with that of the Rodin and the Rodin, Herman, and Schachter studies strongly supports the hypothesis that the responsiveness exhibited by the obese to external food cues is only one aspect of a more general responsiveness to external stimuli.

It is also the case that this "general externality" notion is a powerful predictor of behavior only if a measure of the salience of the external stimulus is taken into account. It is only to stimuli high in salience that the obese are more responsive than normals; when presented with stimuli low in salience, they are no more responsive. In fact, in the first two studies described in this chapter, obese subjects appeared to be less responsive than normals to stimuli low in salience.

In summary, it appears that differences in eating behavior between obese and normal subjects may be a special case of a much broader individual-difference variable. The notion of obese-normal differences in general responsiveness or sensitivity to salient cues in the environment is an intriguing one. Additionally, it holds the appeal of subsuming under one concept a large number of different aspects of behavior from eating to emotionality to performance on cognitive tasks. However, now that differences in such a general trait have been demonstrated, much work is needed to clarify the underlying process mediating such differences.

APPENDIX

This appendix presents details on the numerous studies on which we have based our conclusions about the parallels between hyperphagic animals and obese humans. We have attempted in this section to describe the relevant details of each experiment cited, to identify precisely the data on which our analyses and conclusions were based, and to describe the results of studies which, though possibly relevant, were, for one reason or another, not included in the summary or "batting average" tables in Chapters 1 and 11. In addition, we have attempted to summarize the animal research literature on caloric regulation, emotionality, active and passive avoidance, and the effects of work on food consumption—all of the areas in which we attempted to test on humans findings reported for VMH-lesioned animals.

The appendix is divided into a set of 12 tables for animal studies and 8 for human studies. In the animal tables, studies of static obese animals are always listed first, alphabetically, followed by studies that only tested dynamic hyperphagics. In the tables of Chapter 1, the "batting-averages" and O/N ratios were based only on studies presenting data for already obese animals in the static stage of hyperphagia. This procedure was followed in order to draw the closest possible parallel to our currently obese human subjects. However, for possible usefulness to the reader, experiments using only dynamic animals are also described in the Appendix. Similarly, in Chapter 11, our criterion for inclusion in the analyses required that both dynamic and static animals were tested in a single study. Again, in these Appendix tables, we summarize the results of studies which, though not satisfying this criterion, are possibly relevant.

After each table there is a section which (*a*) provides a description of the procedure of each of the experiments cited and (*b*) specifies, for each experiment, the source of the data reported in the Appendix tables. In the discussion of each experiment, the specific tables, figures, and pages cited refer to the published report of the experiment.

The numbers of the first nine appendix tables are keyed so as to correspond to the tables in Chapters 1 and 11. The following tables are included:

1. The effects of good tasting food on consumption (animals and humans)
2. The effects of bad tasting food on consumption (animals and humans)

3. Amount of food eaten *ad lib* (animals and humans)
4. Number of meals per day (animals and humans)
5. Amount eaten per meal (animals and humans)
6. Speed of eating (animals and humans)
7. The effects of work on food consumption (animals)
8. Activity (animals and humans)
9. Emotionality (animals)
10. Active avoidance (animals)
11. Passive avoidance (animals)
12. The effects of caloric dilution on intake (animals)

In the studies summarized in these tables, if the experimental animals were not specifically labeled as dynamic or static hyperphagics, we considered as static those animals whose weight had reached an asymptote. In some studies, a weight plateau was reached as early as 35 days after surgery; in others, as late as 70 days. The dynamic-static distinction was further complicated by the fact that several studies used different kinds of food during the course of the experiment. These variations typically caused a further weight increase or decrease in animals who had achieved asymptotic levels. Because of these problems, we have tried, in describing each experiment, to be completely explicit about our assignment of animals to the dynamic or static category.

TABLE 1—ANIMAL

The Effects of Good Tasting Food on Consumption

Source	Test Food	Data Reported in:	Amount of Food Eaten By: Normal	Dynamic	Static Obese	Nature of Lesion	Species	Sex	D/N	O/N
Carlisle & Stellar 1969	25% Crisco oil mixed with powdered diet	gms	23 (4)[a]		28 (4)	electro-lytic	rats	male		1.22[b]
	25% isocaloric oil mixed with powdered diet	gms	27 (4)		37 (4)					1.37
Corbit & Stellar 1964	33% vegetable oil with 67% powdered diet	Kcals per day	74 (4)	110.4 (5)		electro-lytic	rats	female	1.49	
	1 gm vegetable oil per 1.5 gm mineral oil mixed with powdered diet		66 (4)		93 (5)					1.41
Graff & Stellar 1962	50% dextrose mixed with powdered diet	gms	16 (13)	18 (8)	27 (6)	electro-lytic	rats	female	1.13	1.69
Lipton 1969	30% lard mixed with ground chow	Kcals	79.20 (5)	153.25 (6)		electro-lytic	rats	female	1.93	

TABLE 1—ANIMAL (Continued)

Source	Test Food	Data Reported in:	Amount of Food Eaten By: Normal	Dynamic	Static Obese	Nature of Lesion	Species	Sex	D/N	O/N
	50% lard mixed with ground chow		86.30 (5)		102.15 (6)					1.18
Miller et al., 1950	60% high fat synthetic diet	gms	15.1 (10)		26.4 (12)	electro-lytic	rats	male		1.75
Teitelbaum 1955	50% dextrose mixed with powdered diet	gms	14.20 (7)	26.52 (5)	29.28 (5)	electro-lytic	rats	female	1.87	2.06
Gold 1970	33% Crisco oil mixed with powdered diet	gms	12.47 (14)	21.03 (19)		bilateral para-sagital knife cuts	rats	male & female	1.69	
Sclafani et al., 1970	33% dextrose mixed with powdered diet	gms	15.30 (5)	35.96 (4)		electro-lytic	rats	female	2.35	
Sclafani & Grossman 1969	25% dextrose mixed with powdered diet	gms	18.03 (5)	33.43 (10)		knife cut	rats	female	1.86	

The *N* for each treatment condition is always reported in brackets under the data for that group.

The data for the two food mixtures are combined with reported in Table 1, Chapter 1.

DESCRIPTION OF ANIMAL STUDIES REPORTED IN TABLE 1

1. The Effects of Good Tasting Food on Consumption

Carlisle and Stellar (1969) mixed three oils with powdered chow in concentrations of 5, 15, 25, 35, and 45% by weight, varying the caloric density and oiliness of the diets. To assess the palatability of the various dilutions, Carlisle and Stellar used the method of paired comparisons, testing mixtures containing 25% Crisco oil, 25% mineral oil, and 25% isocaloric oil. From the data in Figure 2 (p. 111) it is clear that the 25% Crisco oil mixture was highly preferred by lesioned and control animals. Appendix Table 1 compares the relative intake of both groups on the same mixture of powdered diet and 25% Crisco oil offered *ad lib.* The numbers are estimated from data points given in Figure 1 (p. 109) for the 25% Crisco oil mix. Data are also given in the Appendix table for the 25% isocaloric oil mixture which was the second most preferred dilution.

It is unclear whether lesioned animals should be designated dynamic or static. Testing began 2 months postoperatively when hyperphagic animals had already attained a mean body weight of 581 grams. Their *ad lib* pellet intake immediately prior to testing (estimated from Figure 1) was only slightly higher than that of unoperated controls. For these reasons they appear to be static. Yet, Carlisle and Stellar report that lesioned animals gained a mean of 200 grams from the start to the end of testing as compared to a mean gain of 77 grams for the controls. These data suggest that all or some lesioned animals may still have been in the dynamic phase of hyperphagia during testing or at least underwent a new dynamic stage when given good-tasting food.

Corbit and Stellar (1964) gave normal-weight animals and hyperphagics, reduced to their preoperative weights, four diets differing in taste and texture. On each successive diet, animals were fed *ad lib* until their body weight reached asymptote, i.e., no net change in body weight for at least 5 days. When this criterion was met, a new diet was offered. The order of presentation of test diets was as follows: powder, pellets, high fat, pellets, mineral oil.

Although the authors do not differentiate between dynamic and static animals, we have employed the following judgment rule to make this distinction. Body weights and intake for all treatments were reported in Figure 1 (p. 65). Since hyperphagic animals showed a large increase in weight on the pellet diet which directly preceded the high fat diet, lesioned animals were called dynamic during the high fat test. Prior to the mineral oil test, both hyperphagics and controls showed identical patterns of small weight loss and some stabilization. Since no further weight gain was evident, lesioned animals tested on the mineral oil diet were called statics. The data presented in Appendix Table 1 were estimated from Figure 1 (p. 65).

Graff and Stellar (1962) used powdered food mixed with 50% dextrose presented *ad lib* for three consecutive 24-hour periods. The data were estimated from points given in Figure 1 (p. 419) for groups I (obese hyperphagics), II (nonobese hyperphagics who showed a final postoperative mean weight of 100 grams greater than unoperated animals and are therefore considered dynamic during testing), and V (unoperated controls).

Lipton (1969) gave *ad lib* high fat diet (30% lard, by weight, mixed with 70% ground Purina chow, or 50% lard and 50% chow) to lesioned and control animals. After 30 days (a 10 day recovery period and a 20 day "postoperative" period), all animals were given the 30% fat diet for 20 days, followed by the 50% fat diet for 10 days. From the weights reported in Table 2 (p. 511), lesioned animals were considered to be in the dynamic phase of hyperphagia during the 30% fat test session since their weights continued to increase sizably over this period. Since, during the 50% fat diet period their weights reached an asymptotic level, the animals were considered to have attained the static stage of hyperphagia. Data, reported as means of 5-day periods for food intake in Kcals, were taken from Table 2 and pooled across each type of diet. Two means for control animals, one taken from the pooled means for the 30% fat diet test period and the other for the 50% fat diet session, are reported in the Appendix table.

Miller et al. (1950) fed experimental and control animals a 60% high fat synthetic diet on which obese hyperphagics gained considerably more weight than did normals. Appendix Table 1 gives the data reported on page 258 for the high fat diet with a low quinine adulteration.

TABLE 1—HUMAN

The Effects of Good Tasting Food on Consumption

Source	Test Food	Data reported in:	Amount eaten by: Normal	Obese	Sex	O/N
Decke 1971	milk shake	oz	10.55 (9)	13.90 (5)	male	1.32
Nisbett 1968b	vanilla ice cream	gms	152.9 (56)	230.4 (28)	male	1.51
Nisbett & Gurwitz 1970	Enfamil	cc	58 (28)	74 (14)	male & female babies	1.28

Lesioned animals are considered static since their body weights, as indicated in Figure 1 (p. 257), had reached asymptote before the high fat diet was introduced.

Teitelbaum (1955) mixed 50% dextrose by weight with a standard powdered diet to produce a good-tasting flavored diet. Data were taken from Teitelbaum's Table 3 (p. 160) which reports mean food intake in grams for obese, dynamic, and normal animals on the dextrose mix.

Gold (1970) used bilateral parasagittal knife cuts to produce hypothalamic hyperphagic animals. Lesioned and sham-operated animals were given *ad lib* access to a high fat diet consisting of 33% melted Crisco shortening and 67% laboratory meal for seven weeks following surgery. Lesioned animals continued to gain weight steadily over the 7-week period (Figure 1, p. 349) and were therefore considered dynamic. The food intake data, estimated from Figure 1 (p. 349), were pooled for males and females and combined over the means reported for the 7 postoperative weeks.

Sclafani et al. (1970) measured intake of *ad lib* powdered food adulterated with 33.3% dextrose for dynamic hyperphagic and control animals. The % changes in food intake were reported in Table 2, p. 399. From these figures we estimated the actual *ad lib* intake [baseline ad lib intake on unadulterated diet + (% change after adulteration × baseline intake)], in order to make the data more comparable to those reported in other studies. Since testing began in the sixth postoperative week, the data reported in Figure 3 (p. 398) for the last (9th) four-day average were used as the *ad lib* intake baseline. Lesioned animals were considered dynamic since weight curves for VMH rats (Figure 3, p. 398) show a steadily increasing function. Data for animals with amygdaloid lesions are not reported here.

Sclafani and Grossman (1969) fed powdered food mixed with 25% dextrose for two consecutive days *ad lib* to dynamic hyperphagic (2 weeks postoperatively) and normal animals. The data were reported in Table 1 (p. 536) in terms of % change and were converted, for present purposes, to absolute amount eaten [baseline *ad lib* intake on unadulterated diet + (% change after adulteration × baseline intake)]. Data were pooled for hyperphagic animals designated "cut" or "lesioned" by the authors.

DESCRIPTION OF HUMAN STUDIES REPORTED IN TABLE 1

1. The Effects of Good Tasting Food on Consumption

Decke (cited in Schachter, 1971) gave obese and normal-weight adult males a rich, good tasting milkshake which they were to judge for taste. *S*s were given a 40 oz. container and told to drink as much or as little as they wished.

TABLE 2–ANIMAL

The Effects of Bad Tasting Food on Consumption

Source	Test Food	Data reported in:	Amount eaten by: Normal	Dynamic	Static obese	Nature of lesion	Species	Sex	D/N	O/N
Graff & Stellar 1962	powdered diet with .10% quinine	gms	13 (13)	3 (8)	12 (6)	electro-lytic	rats	female	0.23	0.92
Hamilton & Brobeck 1964	pellets dipped in quinine	gms	141.7 (3)	235.5 (2)	232.0 (3)	electro-lytic	monkeys	male & female	1.66	1.64
Miller et al., 1950	hi fat chow with 1.024% quinine	gms	8.8 (10)		2.8 (11)	electro-lytic	rats	male		0.32
Teitelbaum 1955	powdered diet with .125% quinine	gms	15.8 (7)	24.36 (5)	2.14 (7)	electro-lytic	rats	female	1.54	0.14
Sclafani et al., 1970	powdered diet with .10% quinine	gms	10.54 (5)	14.2 (4)		electro-lytic	rats	female	1.34	
Sclafani & Grossman 1969	powdered diet with .10% quinine	gms	10.66 (5)	14.97 (10)		knife cuts	rats	female	1.40	

TABLE 2—HUMAN

The Effects of Bad Tasting Food on Consumption

Source	Test food	Data reported in:	Amount eaten by: Normal	Obese	Sex	O/N
Decke 1971	milk shake with .04 gm quinine per quart	oz	6.4 (9)	2.6 (5)	male	0.40
Nisbett 1968b	ice cream with 2.5 gm quinine per quart	gms	46.8 (56)	59.8 (28)	male	1.28

Nisbett (1968b) gave French vanilla ice cream in a "taste" experiment to obese and normal undergraduate males. *S*s were provided with a quart of ice cream and told to eat as much as they liked until the *E* returned 10 minutes later. The data were taken from Table 2 (p. 110) for the good-tasting ice cream, pooling both deprivation conditions. Underweight and normal *S*s were combined.

Nisbett and Gurwitz (1970) gave heavy and normal infants 120 cc of the standard hospital formula adulterated with 4.7% sucrose solution to make a sweetened version of the formula. Nurses were instructed to let the infants have as much of the sweetened formula as they wanted. The data were estimated from Figure 1 (p. 247), comparing heavy infants to medium and light infants combined. Results for male and female *S*s were pooled.

DESCRIPTION OF ANIMAL STUDIES REPORTED IN TABLE 2

2. The Effects of Bad Tasting Food on Consumption

Graff and Stellar (1962) used powdered food adulterated with 0.1% quinine, given *ad lib* for three consecutive 24-hour periods, followed by 3 days in which plain, powdered food was presented. Data were estimated from bar graphs in Figure 1 (p. 419) for obese and nonobese hyperphagics and unoperated controls.

Hamilton and Brobeck (1964), measuring the effects of palatability on the intake of hyperphagic monkeys, added quinine to monkey chow pellets by dipping pellets in varying quinine solutions (1 gm quinine per liter water, or 2, 4, or 8 gms per liter) for 10 seconds. On test days approximately one week apart, pellets with increasing quinine concentrations were presented in place of regular chow. The data presented in Appendix Table 2 were obtained by averaging the means for each group of animals (dynamic, obese, and controls) across all quinine solutions reported in Table 1 (p. 278).

Miller et al. (1950) added quinine to a synthetic, high fat diet from the 61st to the 70th day after the operation. Data in Appendix Table 2 represent intake on day 70 for a mixture of 1024 mg. quinine per 100 gms food (reported on p. 258 of Miller et al.).

Teitelbaum (1955) combined .125% quinine by weight with a standard powdered diet and provided the mixture *ad lib* to static, dynamic, and control animals. Data were taken from Table 2, p. 159.

Sclafani et al. (1970) gave powdered diet adulterated with 0.1% quinine *ad lib* in the home cage to hyperphagic and control animals beginning in the sixth postoperative week. (See discussion of Sclafani et al. in "effects of good taste" section for description of how data reported here were obtained.)

Sclafani and Grossman (1969) measured intake of *ad lib* powdered food mixed with 0.1% quinine in dynamic hyperphagic (2 weeks postoperatively) and control animals. (See discussion of Sclafani & Grossman, 1969, in "effects of good taste" for description of how data reported here were obtained.) Data for "cut" and "lesion" groups were pooled.

DESCRIPTION OF HUMAN STUDIES REPORTED IN TABLE 2

2. The Effects of Bad Tasting Food on Consumption

Decke (cited in Schachter, 1971) added .04 gm quinine per quart of milkshake. The same *S*s who were tested for consumption of the good-tasting milkshake participated in this study. The good and bad milkshakes were presented in counterbalanced order.

Nisbett (1968b) adulterated vanilla ice cream with 2.5 gm quinine/quart of ice cream. Data were taken from Table 2 (p. 110) for the bad-tasting mix. Data for normal and underweight *S*s were combined.

DESCRIPTION OF ANIMAL STUDIES REPORTED IN TABLE 3

3. Amount of Food Eaten *Ad Lib*

Brooks et al. (1946) provided rats with an *ad lib* diet of wet mash before and after lesioning. The data are taken from Table 1 of Brooks et al. (p. 738) which reports data which is apparently representative of that obtained from a somewhat larger sample of unspecified number.

Corbit (1965) measured the amount of pellet food eaten per day by lesioned obese and normal control animals before and after a test for quinine finickiness. The data were taken from Table 1 (p. 124) and pooled for days 1, 2, 8, 9, and 10 when food and water were available *ad lib.*

Ferguson and Keesey (1971) measured mean *ad lib* intake of dry ground Purina chow for hyperphagic and normal animals. On the basis of body-weight data given in Figure 4 (p. 268), the E-1 group (who had been fed *ad lib* throughout the post-lesion period) appeared to have attained an asymptotic body weight by day 19 after the operation and were therefore considered static hyperphagics for the present analysis. The data were taken from Table 2 (p. 268) for E-1 and control *S*s on postlesion days 19-21. Animals in the E-1 group were considered dynamic on days 1-18.

Graff and Stellar (1962) in Figure 1 (p. 419) report intake data for animals fed *ad lib* on a standard powdered diet for one week prior to finickiness tests. We include data in Appendix Table 3 for Graff and Stellar's groups I, II, and V—the obese and nonobese hyperphagics, and unoperated controls. However, it is unclear whether the nonobese hyperphagics (Group II) should be called dynamic since they never became obese.

Hamilton and Brobeck (1964) fed animals standard pellets and their average daily intake was measured. The data were taken from four-day pretest measures of ingestion prior to quinine rejection tests reported in Table 1 (p. 278).

Hetherington and Ranson (1942) measured the number of grams of food eaten per day by hyperphagic rats and littermate controls. All animals received *ad lib* pellets or wet mash. All lesioned animals attaining adiposity [judged to be from + to +++ in degree of obesity in Table 1 (p. 611)] were considered static hyperphagics. Because the obesity determination was made at the end of the experiment, we only report data in Appendix Table 3 for the fifth to the final postoperative week after the food intake of lesioned animals appeared to reach an asymptote. No actual weight curves were provided.

Nachman (1967) measured *ad lib* pellet intake over 3 consecutive days when water was also continuously available. Data for mean intake for obese hyperphagic and control animals (tested 30-60 days after the operations) were reported in the text on p. 127.

Sclafani (1971) measured *ad lib* intake of a pellet diet for 21 consecutive days (8-10 weeks after surgery) in operated and control animals. The three experimental groups which showed hyperphagia were animals with 1) (VMH) electrolytic lesions, 2) (VL-1) long lateral parasaggital cuts, and 3) (VL-2) either long parasaggital cuts slightly posterior to those of VL-1 or short cuts rostral to another group (VL-4) which did not become hyperphagic. The data reported in Appendix Table 3 are for the 20 and 21st days when according to the weight chart (Fig. 7, p. 77), hyperphagic *S*s were in the static stage. The data for food intake are taken from Table 1 (p. 78), using the "term" column which refers to terminal measures, Days 20 and 21, when animals were on the pellet diet. Data for *S*s with parasaggital transections (VL-1 and VL-2) were combined.

TABLE 3–ANIMAL

Amount of Food Eaten *Ad Lib*

Source	Test food	Data Reported in:	Amount eaten by: Normal	Dynamic	Static obese	Nature of lesion	Species	Sex	D/N	O/N
Brooks et al., 1946	Purina chow with equal weight water	gms	36.5 (5)	64.4 (5)	41.4 (5)	electro-lytic	rats	?	1.76	1.13
Corbit 1965	pellets	gms	17.0 (4)		22.5 (5)	electro-lytic	rats	female		1.32
Ferguson & Keesey 1971	dry ground Purina chow	gms	18.2 (4)	28.98 (8)	21.6 (8)	electro-lytic	rats	female	1.59	1.19
Graff & Stellar 1962	powdered chow	gms	16 (13)	15 (8)	22 (6)	electro-lytic	rats	female	0.94	1.38
Hamilton & Brobeck 1964	pellets	gms	190 (3)	310 (2)	250 (3)	electro-lytic	monkeys	male & female	1.63	1.32
Hetherington & Ranson 1942	pellets, mash, special diet	gms	15.5 (6)		16.7 (6)	electro-lytic	rats	male		1.08
Nachman 1967	pellets	gms	16.3 (10)		23.4 (11)	electro-lytic	rats	female		1.44

TABLE 3—ANIMAL (Continued)

Source	Test food	Data Reported in:	Amount eaten by: Normal	Dynamic	Static obese	Nature of lesion	Species	Sex	D/N	O/N
Sclafani 1971	pellets	gms	18 (12)		32 (13)	electro-lytic	rats	female		1.78
	pellets	gms	18 (12)		33 (21)	ventro-lateral knife cuts	rats	female		1.83
Teitelbaum 1955	powdered chow	gms	16.48 (7–14)	25.96 (5–10)	20.63 (5–12)	electro-lytic	rats	female	1.58	1.25
Teitelbaum & Campbell 1958	pellets	number	323.1 (6)	625.7 (7)	344.9 (4)	electro-lytic	rats	female	1.94	1.07
	liquid	ml	42.7 (5)	87.1 (6)	45.3 (5)	electro-lytic	rats	female	2.04	1.06
Balagura & Devenport 1970	pellets	number	221.6 (9)	485.5 (9)		electro-lytic	rats	male & female	2.19	
Corbit & Stellar 1964	pellets	Kcals	62 (4)	100 (5)		electro-lytic	rats	female	1.61	
Thomas & Mayer 1968	liquid diet diluted with 50% water	cals	72 (6)	151 (6)		electro-lytic	rats	female	2.10	
Williams & Teitelbaum 1959	liquid diet	ml	45 (6)	77 (3)		electro-lytic	rats	female	1.71	

Teitelbaum (1955) established a baseline level of consumption for each experimental group fed *ad lib* on a standard powdered diet during a five-day control period. Data in Appendix Table 3 were obtained by pooling the data (for animals fed the unadulterated diet) given in Teitelbaum's Tables 2 and 3 (p. 159 and 160). It is unclear, however, how many of the same animals were used in each experiment, so we have reported the range of the number of subjects possibly included in the data for each treatment condition.

Teitelbaum and Campbell (1958) housed animals for seven days in Skinner boxes where a single bar press always produced one pellet of food. The number of presses per day is used as an index of daily consumption. Data represent the mean amount eaten per day over the last five test days. The authors also studied ingestion patterns for liquid diet. The liquid food was given *ad lib* by means of a drinking tube fixed to the front of the cage. Each lick was automatically recorded and, in addition, volume measurements were taken at 12-hour intervals. The data were given in Table 1 (p. 137) for the last five of seven test days.

Balagura and Devenport (1970) measured daily ingestion of animals required to press a bar in order to receive a pellet. The same animals were tested before and soon after lesions were made. Consequently, hyperphagic animals are considered dynamic. The data were estimated from Figure 3 (p. 361) by multiplying the number of meals eaten per day by the average meal size for control and lesioned animals. Data for males and females are pooled.

Corbit and Stellar (1964) measured the *ad lib* intake of a standard pellet diet for dynamic hyperphagics and normal controls. Hyperphagic animals were judged to be dynamic since they had been starved to normal weight and were in a phase of rapid weight gain when the pellet test was given. The test period varied between 15 and 70 days, depending on the number of days necessary for animals to reach asymptotic body weight. The data were estimated from the second column in Figure 1 (p. 65) for the first pellet test. Data from the second pellet test, occurring after a high fat diet, appear to be greatly affected by having followed the preferred high fat diet and therefore were excluded from the present analysis.

Thomas and Mayer (1968), measuring regulation of food intake, gave liquid diet diluted with 50% water to normal and hyperphagic animals *ad lib.* Since the experimental sessions began after a 10-day postoperative recovery period, operated animals were considered to be in the dynamic stage of hyperphagia. Data were based on means from a 4-day test period and were reported on p. 645. Animals received a fixed amount of liquid diet for each bar press and the number and pattern of presses were recorded. Data were given in calories.

Williams and Teitelbaum (1959)* assessed the prerequisites for recovery from the lateral hypothalamic syndrome, using control groups of dynamic hyperphagics and unoperated animals. Data reported in Appendix Table 3 were estimated from Figure 1 (p. 460) and averaged for test days 1-3 when a full strength liquid diet was provided to the dilution manipulation on day 4.

*In addition to the studies cited here, several more recent articles also report *ad lib* intake data for dynamic hyperphagics (Cox, Kakolewski, & Valenstein, 1969; Lipton, 1969; Sclafani & Grossman, 1969, 1971; Sclafani et al., 1970; Smith & Britt, 1971; Palka, Liebelt, & Critchlow, 1971). Since all of them report that dynamic hyperphagic rats eat more than controls, we decided that a further cataloguing of supportive data was superfluous.

TABLE 3—HUMAN

Amount of Food Eaten *Ad Lib*

Source	Test food	Data reported in:	Amount eaten by: Normal	Amount eaten by: Obese	Sex	O/N	Source of data
Ross Chapter 7	nuts	gms	22.5 (60)	27.9 (60)	male	1.24	
Schachter & Friedman Chapter 2	nuts	gms	7.43 (20)	9.18 (20)	male	1.24	
Schachter & Gross 1968	crackers	gms	28.8 (24)	28.8 (22)	male	1.00	
SELF-REPORT							
Beaudoin & Mayer 1953	normal daily diet	cals	2198 (58)	1964 (59)	female	0.89	1-day record
	normal daily diet	cals	2196 (12)	1591 (12)	female	0.72	3-day record
	normal daily diet	cals	2201 (20)	2829 (33)	female	1.29	research dietary history
Johnson et al., 1956	normal daily diet	cals	2706 (28)	1965 (28)	female	0.73	
Ross et al. 1971	normal daily diet	gms	1731 (62)	1819 (14)	male	1.05	
Stefanik et al., 1959	normal daily diet	cals	4052 (14)	3221 (14)	male	0.80	

DESCRIPTION OF HUMAN STUDIES REPORTED IN TABLE 3

Amount of Food Eaten *Ad Lib*

Behavioral

Although the average daily consumption for obese and normal human *S*s has not been measured directly, it seems plausible to assume that those studies where *S*s were allowed *ad lib* access to food during the experimental session are comparable to animal studies of *ad lib* feeding. In all such experiments on humans, eating is placed in an *ad lib* context; that is, a bowl of ordinary food, such as nuts or crackers, is placed in a room, the experiment presumably has nothing to do with eating, and the subject is free to eat or not as he chooses, just as is a rat in its cage. In all of these studies overweight *S*s were at least 15% heavier than the averages set by the Metropolitan Life Insurance norms (1959). All *S*s were male undergraduates.

Ross (Chapter 7) measured the number of grams of nuts eaten from a large tin of cashew nuts on a table in front of each *S*. Data were taken from Table 1 reporting the mean consumption across all treatment groups during the first experimental session.

Schachter and Friedman (Chapter 2) gave subjects free access to either shelled or unshelled almonds and measured the number of grams eaten by obese and normal *S*s who were filling out personality questionnaires. The data reported in Appendix Table 3 were pooled for all experimental conditions.

Schachter and Gross (1968) allowed subjects free access to crackers while they were supposedly participating in a study correlating physiological and psychological variables. Data were taken from Table 2 (p. 101) poooling "time" treatment conditions for both weight groups.

Self Report

Beaudoin and Mayer (1953) used several assessment methods to study adult female populations for whom height and weight records were available in order to estimate average daily caloric intake. The data reported in Table 1 (p. 30) give figures for *S*s who completed a one-day food record during midweek, *S*s who completed a midweek 3-day intake record, and *S*s for whom dietary histories were taken by Public Health hospital staff members. We report these data separately in Appendix Table 3 since they represent different samples.

Johnson et al. (1956) selected a sample of obese and normal girls from the upper grades of high school. The obese girls were chosen on the basis of having maintained, for one year, a heavy physique status (A4 or greater) according to a modified Wetzel Grid. Interviews were used to provide dietary information. The average daily caloric intakes were reported in Table 1, p. 40.

Ross et al. (1971) surveyed the eating habits of Columbia undergraduates for 13 consecutive days. During that time, *S*s recorded everything they ate or drank. Data, given in mean number of grams eaten per day, were reported for food having some caloric value. Data were also reported for mean number of calories per day. Obesity was determined by degree of overweight on Metropolitan Life Insurance charts (1959). *S*s 15% or more overweight for a given height and age were considered obese. The remaining *S*s, considered normal, ranged from 27.3% underweight to 14.9% overweight.

Stefanik et al. (1959) used dietary interviews with adolescent boys to determine average daily food intake for the school year and for an 8-week summer camp season. *S*s estimated their average intake for the school year during the first week of the camp season and estimated their average intake for the 8 weeks at camp during the last week of the camp season. Obesity was determined by skinfold measurement. Data for mean daily food intake during school year (Table 4, p. 58) and during the camp season (Table 5, p. 59) were pooled for the present analysis.

TABLE 4—ANIMAL

Number of Meals Per Day

Source	Test food	Number meals eaten by: Normal	Dynamic	Static obese	Nature of lesion	Species	Sex	D/N	O/N
Brooks, et al., 1946	Purina chow with equal water weight	10.2 (5)	9.2 (5)	9.6 (5)	electro-lytic	rats	?	0.90	0.94
Larsson & Strom 1957	standard diet (reported in range of number of meals)	Light: 35-45 Heavy: 20-25		9–15	GTG	mice	male		
Teitelbaum & Campbell 1958	solid pellets	11.0 (6)	13.3 (7)	8.3 (4)	electro-lytic	rats	female	1.21	0.75[a]
	liquid	11.5 (5)	11.1 (6)	9.9 (5)	electro-lytic	rats	female	0.97	0.86
Balagura & Devenport 1970	pellets	5.26 (9)	7.60 (9)		electro-lytic	rats	male & female	1.44	
Thomas & Mayer 1968	liquid diluted with water	16.2 (6)	11.2 (6)		electro-lytic	rats	female	0.69	

[a] The data for pellets and liquid are combined when reported in Table 4, Chapter 1.

TABLE 4—HUMAN
Number of Meals Per Day (Self Report)

Source	Test food	Number meals eaten by: Normal	Obese	Sex	O/N
Beaudoin & Mayer 1953	normal daily diet	4.36 (58)	3.98 (59)	female	0.91
Johnson et al. 1956	see accompanying text				
Ross et al. 1971	normal daily diet	4.29 (62)	3.99 (14)	male	0.93

DESCRIPTION OF ANIMAL STUDIES REPORTED IN TABLE 4

Number of Meals Per Day

Brooks et al. (1946). Meal-frequency data for five animals preoperatively and in each stage of hyperphagia were reported in Table 1 (p. 738).

Larsson and Strom (1957), investigating the characteristics of goldthioglucose obesity in mice, measured *ad lib* ingestion patterns. Although the actual means for each group were not given, the authors reported the ranges for experimental and control groups. Normal animals were further divided into heavy (weighing 30-40 gms) and light (weighing 20-25 gms) groups. The authors reported data only for those hyperphagic animals with fully developed obesity.

Teitelbaum and Campbell (1958). The data for liquid diet were taken from Table 1 (p. 137) and for solid diet from Table 2 (p. 139), for control animals and dynamic and static hyperphagics.

Balagura and Devenport (1970). Male and female rats, tested pre- and postoperatively, were measured for the number of meals of laboratory pellets eaten each day. The data were taken from Figure 3 (p. 361), pooling males and females.

Thomas and Mayer (1968). Data, for liquid diet, were taken from means reported on p. 644.

DESCRIPTION OF HUMAN STUDIES REPORTED IN TABLE 4

Number of Meals Per Day (Self Report)

Beaudoin and Mayer (1953) reported the percentage of *S*s eating a particular meal or snack in Table 3 (p. 31). Mean number of eating instances was obtained by multiplying these percentage figures by the number of *S*s in each condition, summing the obtained figures, and dividing by the total number of subjects in each category.

Johnson et al. (1956) reported snacking patterns and breakfast data for obese and normal *S*s on p. 40. Although it is impossible to compute the actual data, given the snacking patterns plus the data Mayer's group has reported in similar studies about the likelihood of eating lunch and dinner, it seems safe to conclude that overall, obese *S*s ate less frequently than did normals.

Ross et al. (1971) calculated the mean number of times *S*s reported eating a full meal or snack each day.

TABLE 5–ANIMAL

Amount Eaten Per Meal

Source	Test food	Data reported in:	Amount eaten by: Normal	Dynamic	Static obese	Nature of lesion	Species	Sex	D/N	O/N
Brooks et al., 1946	wet mash	gms	3.6 (5)	7.5 (5)	4.2 (5)	electro-lytic	rats	?	2.08	1.17
Teitelbaum & Campbell 1958	liquid diet	ml	4.1 (5)	8.2 (6)	4.9 (5)	electro-lytic	rats	female	2.00	1.20
	pellets	number	30.8 (6)	50.9 (7)	46.3 (4)	electro-lytic	rats	female	1.65	1.50
Balagura & Devenport 1970	pellets	number	48.7 (9)	65.5 (9)		electro-lytic	rats	male & female	1.35	
Thomas & Mayer 1968	liquid diet	cals	4.5 (6)	12.4 (6)		electro-lytic	rats	female	2.76	

DESCRIPTION OF ANIMAL STUDIES REPORTED IN TABLE 5

Amount Eaten Per Meal

Brooks et al. (1946). The average amount of wet mash eaten per meal was measured preoperatively and during the dynamic and static phase of hyperphagia in the same animals. The data are reported in Table 1 (p. 738).

Teitelbaum and Campbell (1958) measured ingestion patterns of animals for both liquid and solid foods. In the solid food test, animals received one pellet per bar press. A single meal was defined as "any burst of food intake of at least five pellets separated by at least five minutes from any other burst." The data were taken from Table 2 (p. 139). The authors specified the criterion for a single liquid meal as "a continuous period of drinking which was not interrupted by a pause of more than 5 minutes." Data were taken from Table 1 (p. 137).

Balagura and Devenport (1970) tested the same animals pre- and postoperatively. A meal was defined as "a series of five or more panel-push responses preceded and followed by at least 20 minutes." Animals were given 1 pellet for each response. The data were read from bar graphs presented in Figure 3 (p. 361). In Appendix Table 5 data for males and females were pooled.

Thomas and Mayer (1968) defined a meal as 3 or more reinforced bar presses (yielding about 1 ml. of liquid diet) "separated by less than 3 minutes from each other and by not less than 20 minutes from any previous group." The data for amount eaten per meal were reported, in calories, on p. 644.

TABLE 5—HUMAN
Amount Eaten Per Meal

Source	Test food	Data reported in:	Amount eaten by: Normal	Amount eaten by: Obese	Sex	O/N	
BEHAVIORAL							
Nisbett 1968a	roast beef sandwiches	whole sandwiches	1.74 (48)	1.90 (21)	male	1.09	
Nisbett & Gurwitz 1970	Enfamil formula	cc	54 (28)	58 (14)	male & female babies	1.07	Experiment I
	Enfamil	cc	54 (23)	72 (11)	male & female babies	1.33	Experiment II
Nisbett & Storms in press	tuna, roast beef & turkey sandwiches	quarter sandwiches	1.58 (80)	2.48 (34)	male	1.57	Experiment II
	tuna sandwiches	gms	143.75 (24)	171.25 (24)		1.19	Experiment III
Pliner Chapter 5	roast beef & chicken sandwiches	quarter sandwiches	6.24 (48)	7.15 (48)	male	1.15	
SELF REPORT							
Beaudoin & Mayer 1953	normal daily diet	cals	503.9 (58)	495.33 (59)	female	0.98	
Ross et al. 1971	normal daily diet	gms	400 (62)	456 (14)	male	1.14	

DESCRIPTION OF HUMAN STUDIES REPORTED IN TABLE 5

Amount Eaten Per Meal (BEHAVIORAL)

In all of the studies reported below, subjects were given a real meal. They had access to more food than they could possibly eat and they were encouraged to eat as much as they wished. Eating was terminated whenever *S*s chose to stop. The only data included in the Appendix table are those where there was an actual meal provided for the *S*. In some experiments, there were eating instances of strange food where *S*s ate nothing but milkshakes, ice cream, or crackers. Although they cannot be taken to reflect amount eaten at "real" meals, the data reported below generally support the meal-size data.

Study	N	O	O/N	Kind of food	Units
(a) Schachter et al. (1968)	18.1	18.3	1.01	crackers	number eaten
(b) Goldman (1968) averaged over all sessions and treatment conditions	1117.5	1125.5	1.01	milkshake	gms
(c) Nisbett (1968b)	99.85	145.1	1.45	ice cream	gms

Obese *S*s were 15% or more overweight as determined by the Metropolitan Life Insuranc norms (1959). All *S*s were male undergraduates.

Nisbett (1968a) offered *S*s either one or three visible road beef sandwiches and announced that there were dozens more in the refrigerator. *S*s were told that this meal was intended to replace the lunch they had skipped in order to participate in the experiment. Data were taken from Table 1 (p. 1254) and averaged over all experimental conditions. Means for underweight and normal *S*s were pooled.

Nisbett and Gurwitz (1970)

(*a*) Experiment I: The intake of heavy and light new-born infants on the hospital's standard Enfamil formula are reported for 1-4 days after birth. The data were estimated from Figure 1 (p. 247) for the unsweetened diet, pooling medium and light infants. Grouping into weight categories was done by the experimenters on the basis of weight at birth. Data for male and female *S*s are combined in Appendix Table 5.

(*b*) Experiment II: Infants were divided into overweight and normal groups on the basis of the formula: weight/length^3. For feedings 1 and 3 on two consecutive days, all infants were offered unsweetened Enfamil formula through a standard-sized nipple. The data reported in Appendix Table 5 were estimated from Figure 2. Data for medium and small infants were combined.

Nisbett and Storms (in press)

(*a*) Experiment II: *S*s were given a platter of tuna, roast beef, and turkey sandwiches to replace dinner which they had skipped for the experiment. Data given in Table 1 (p. 8) were combined across all treatment conditions to reflect average number of quarter sandwiches eaten by each weight group. Figures were pooled for normal and underweight *S*s.

(*b*) Experiment III: In this study, *S*s were offered a platter of tuna fish sandwiches cut into either quarters or sixteenths. The data, taken from Figures 2A and B were pooled for the present analysis to obtain a mean score for obese and normal *S*s combining the different size of sandwich and deprivation conditions. The data are reported in mean number of grams eaten.

Pliner (Chapter 5) also gave *S*s sandwiches to replace the lunch meal which the experiment required them to skip. Data reported in Appendix Table 5 gave the mean number of quarters eaten for *S*s in solid and liquid preload conditions combined.

TABLE 6—ANIMAL AND HUMAN
Speed of Eating

Source	Test food	Data Reported in:	Normal	Dynamic	Static obese	Nature of lesion	Species	Sex	D/N	O/N
Animal										
Teitelbaum & Campbell 1958	liquid	licks per sec	5.6 (5)	5.2 (6)	5.0 (5)	electro-lytic	rats	female	0.93	0.89
	pellets	# per minute	4.0 (6)	3.4 (7)	5.1 (4)	electro-lytic	rats	female	0.85	1.28
Brobeck et al., 1943	wet mash			report voracious eating postoperatively		electro-lytic	rats		D>N	
Brooks et al., 1946	wet mash			report voracious eating postoperatively		electro-lytic	rats		D>N	
Wheatley 1944	wet mash			report voracious eating postoperatively		electro-lytic	cats		D>N	
Human										
Nisbett 1968b	ice cream	spoons per minute	4.85 (56)		6.12 (28)			male		1.26

SELF-REPORT

Beaudoin and Mayer (1953), in Table 3 (p. 31), reported data from one-day food intake records of mean consumption for each meal and snack eaten. The average number of meals per subject, obtained by multiplying the percentage of *S*s eating each meal by the total number of *S*s, was divided into the total caloric intake (given in Table 1, p. 30) to provide an index of mean consumption per eating instance.

Ross et al. (1971) reported the average amount eaten on each eating occasion (defined as a meal or a snack) by obese and normal *S*s over a 13 day period.

DESCRIPTION OF ANIMAL AND HUMAN STUDIES REPORTED IN TABLE 6

Speed of Eating (Animal)

Teitelbaum and Campbell (1958) determined rate by measuring, on the cumulative record, the slope of each eating burst, i.e., a single meal, and averaging the slopes. Data were taken from Tables 1 (p. 137) and 2(p. 139) for liquid and solid diets.

Several early studies reported "observations" of eating behavior: **Brobeck et al.** (1943) and **Brooks et al.** (1946), using wet mash, reported that lesioned rats showed voracious, rapid eating in the dynamic phase of hyperphagia. It is unclear, however, whether they were referring to the extended stage of dynamic hyperphagia or simply the immediate postoperative period. According to **Wheatley** (1944) recently lesioned cats with ventromedial damage either "ate hurriedly or wolfed their food, while normal cats given the same diet ate slowly, frequently taking several hours to consume their allowance."

Speed of Eating (Human)

Nisbett (1968b) observed the number of spoonfuls of ice cream eaten by *S*s each minute (data not reported in article). From these data, rates of ingestion were determined by calculating the total number of spoonfuls divided by minutes. Figures in Appendix Table 6 reflect pooled means for hungry and full *S*s in both weight groups eating good ice cream.

TABLE 7—ANIMAL

The Effects of Work on Food Consumption

Source	Activity measured	Data measured in:	Normal	Dynamic	Static obese	Nature of lesion	Species	Sex	D/N	O/N	Schedule
Hamilton & Brobeck 1964	bar presses	# presses	150 (3)	400 (2)	300 (2)	electro-lytic	monkeys	male & female	2.67	2.00	FR 1 (easy)
			400	200	50				1.50	0.13	FR 1024 (hard
Teitelbaum 1957	bar presses	# presses	130 (6)	240 (5)	160 (5)	electro-lytic	rats	female	1.85	1.23	FR 1 (easy)
			4500	720	90				0.16	0.02	FR 256 (hard)
Falk 1961	bar presses	differences in number of presses post-op & pre-op		very overweight 2460 not very overweight 394.4		electro-lytic	rats	female			VI 1
Grossman 1966	bar presses	# presses	154 (7)	61 (14)		atropine	rats	male	0.40		FR 1 (easy)
			371	124					0.33		VR 5
Larkin & Kissileff 1971	bar presses	# presses (thousands)	8.6 (3)	14.2 (3)		electro-lytic	rats		1.65		FR 64-2048 (hard)
Marks & Remley 1972	bar presses	responses/min.	29.3 (5)	32 (85% of nor- (4) mal wt)		electro-lytic	rats	female	1.1		VI 2
			29.3 (5)	6.5 (85% of (6) static wt)		electro-lytic	rats	female	0.22		VI 2
Miller et al., 1950	amount of food eaten	gms	16.3 (10)	21.2 (11)		electro-lytic	rats	male	1.30		Without weighted lids (easy)
			10.8	3.3					0.31		weighted lids (hard)

TABLE 7—ANIMAL (Continued)

Source	Activity measured	Data measured in:	Normal	Dynamic	Static obese	Nature of lesion	Species	Sex	D/N	O/N	Schedule
Porter & Allen 1972	bar presses	responses/min.	42.3 (5)	54 (6) (5, 10, 15 & 20% of non-obese wt)		electro-lytic	rats	female	1.28		VI 1
			38 (5)	37.8 (6) (5, 10, 15 & 20% of obese wt)					0.99		
Sclafani 1971	bar presses	# presses	643.75 (8)	456.25 (3)		electro-lytic	rats	female	0.65		FR 1, 2, 4, 8 (easy)
	bar presses	# presses	400 (8)	25 (3)		electro-lytic	rats	female	0.06		FR 256 (hard)
	bar presses	# presses	643.75 (8)	518.75 (5)		ventro-lateral knife cuts	rats	female	0.81		FR 1, 2, 4, 8 (easy)
	bar presses	# presses	400 (8)	650 (5)		ventro-lateral knife cuts	rats	female	1.38		FR 256 (hard)
Singh 1972	bar presses	# presses	218 (16)	300 (19)		electro-lytic	rats	female	1.38		FR 1, 3, 11 (easy)

DESCRIPTION OF ANIMAL STUDIES REPORTED IN TABLE 7

The Effects of Work on Food Consumption

Hamilton and Brobeck (1964) established preoperative bar-press rates on FR schedules of 1, 4, 8, 16, 32, and 64 in one-hour sessions conducted daily with monkeys. Four animals were lesioned and two who developed definite hyperphagia were considered dynamic. Two lesioned animals from an earlier study who had attained obesity were tested as static *S*s. *S*s were tested on FR schedules for 1-hour sessions, first retrained on FRI for 6 days and then FR 4, 8,16, ... , 1024 for 2 days each using banana-flavored pellets. Animals were given 1/5 their pretest daily food intake one hour after testing. The data for number of presses on the easy (FR1) and effortful (FR1024) schedules were obtained from data supplied by Hamilton to N. Mrosovsky (personal communication) but may be estimated from Figure 9 (p. 277).

Teitelbaum (1957) measured food-directed activity in a Skinner box in animals who were 12 hours deprived. After a stable 12-hour food intake level was established, intake was measured for 12-hour periods using FR reinforcement schedules of 1, 4, 16, 64, and 256 given in ascending order for all animals. The data, for mean number of bar presses per 12-hour session, were estimated from Figures 2 and 3 (p. 488) for FR1, a presumably easy schedule, and FR 256, the most effortful.

Falk (1961) held weight of lesioned rats to preoperation levels and measured the difference between post- and preoperation responses on a VI-1 min. reinforcement schedule* for 45 mg food pellets. During postoperative testing, animals were maintained at 75-80% of their starting weights. Testing began the day after surgery. In Figure 1 (p. 67) Falk presents measurements for postoperative number of responses *minus* preoperative number of responses as a function of the weight attained by subjects 12 weeks after testing when on an *ad lib* diet. From this figure, we report two separate means: 1) very obese *S*s (final 12-week weights from 400 to 550 grams) and 2) less obese *S*s (final weights from 280 to 400 grams). All animals were dieted and therefore considered dynamic when testing occurred. Since there was no control group, no comparative D/N scores are possible in Appendix Table 7.

Grossman (1966) directly applied atropine (0.5-5.0 mg) to the ventromedial hypothalamus. Chemically stimulated and control animals were tested for lever pressing in 15-minute sessions under 22 hours of food and water deprivation. Animals were trained until a stable base rate of responding, reached after about 60 days, was attained. Then the daily 15-minute tests were continued under progressively worsening reinforcement schedules up to VR-5. Since animals were not obese at the time of testing, they were placed in the column for dynamic hyperphagics in Appendix Table 7. The data in Appendix Table 7 were estimated from Table 1, p. 5.

Larkin and Kissileff (1971) trained rats to bar press for food on an ascending series of FR schedules: FR 1, 4, 16 , and 64. *S*s received 15 days of continuous training on FR 64 before VMH lesions were made. After surgery, animals were tested again beginning with 5-8 days on FR 64. New, more difficult, schedules not used during pre-lesion training were also tested. These were FR 256 for 3 days and 3 days each on FR 512, 1024, and 2048. Retesting was done immediately following lesioning so that the hyperphagic *S*s were in the dynamic phase. Data for control and lesioned rats at all difficult FR schedules (FR 64-2048) were pooled and estimated from Figure 3.

Marks and Remley (1972) measured bar-pressing rates of VMH lesioned *S*s, produced by direct or radio frequency current, and control *S*s at a VI-2 schedule of reinforcement in pre- and postsurgery testing.* Prior to surgery animals were dieted to 90% of their normal weight and trained on a VI-2 min. schedule. After surgery two groups of operated animals were formed. Hyperphagics at normal weights (Hn) were *S*s which demonstrated hyperphagia following surgery (judged by a 9-day period of weight gain above the weight gain of the heaviest control rat) and were then reduced to 90% of the control rats' weight. Hyperphagics at static weights (Hs) were *S*s which were allowed to attain static weight and then were reduced to 90% of their static

*Most of the other experiments reported here used FR schedules.

weight. "After surgery, and upon reaching the appropriate weight level (Hn or Hs), the rat was returned to the operant box and given a 3 hr. test. The first 2 hrs. consisted of a VI-2 min. schedule followed by a 1 hr. extinction period [p. 99]." In Appendix Table 7 data are reported for the direct current lesioned and control *S*s and are taken from Fig. 1 (p. 101). Since the weight of both operated groups was greatly reduced, we consider all hyperphagic animals to be dynamic in both the Hn and Hs conditions.

Miller et al. (1950) taught lesioned and control animals to obtain high fat synthetic diet by operating light, hinged lids which were put on the food dishes in each rat's home cage. After 6 days of learning to operate the lid, a 75-g. weight was put on the lids for half the lesioned and control animals. After 3 days, the weights were shifted for 3 days to the lids of the remaining ½ of each group. The data were reported on p. 258 for amount eaten without weighted lids (easy conditions) and with weighted lids (effortful). Significance levels were also reported for a measure of strength of pull when animals were temporarily restrained on the way to food in an alley. Animals were still steadily gaining weight and were therefore considered dynamic.

Porter and Allen (1972) measured number of bar presses by VMH lesioned and control *S*s in response to VI-1 min. reinforcement schedule.* Animals were tested in nonobese and obese phases. Nonobese testing began 2 days after surgery at 5, 10, 15, and 20% of preoperation weights. Obese testing began following 51 days of *ad lib* feeding and *S*s were again dieted after this time to 5, 10, 15, and 20% of their obese weight. In both phases . . . "*S*s were tested for 30 minute sessions on the VI-1 schedule. The testing lasted for 4 days . . . Over a 3 day period between each successive weight loss level, *S*s were deprived of food until the correct weight was obtained [p. 286]." Since *S*s underwent weight reduction in both phases, by our criteria they were in the dynamic stages of hyperphagia during all data collection. Data in Appendix Table 7 are taken from the pooled means for hyperphagic *S*s in all weight-reduction conditions. Data for both obese and nonobese phases of testing are reported and are estimated from Fig. 3 (p. 287).

Sclafani (1971) compared bar-press scores for hyperphagic (VMH, VL-1, and VL-2—see Sclafani, 1971, in Appendix Table 3 for exact location of these knife cuts) and control female rats. The VMH rats received electrolytic lesions and the VL-1 and VL-2 received lateral parasagittal transections. The data for the *S*s with parasagittal transections were pooled. Animals were trained at FR-1 in daily 30-min. test periods, and after four days were given three sessions each of FR-2, FR-4, FR-8, FR-16, FR-32, FR-64, FR-128, and FR-256 schedules. All animals were put on a 6 day diet prior to testing to reduce their weight to 85% of the pretest level, and thus all experimental animals were nonobese and considered in the dynamic stage of hyperphagia by our criteria. Data are estimated from Fig. 11 (p. 83), and mean number of lever-press responses are reported for the various FR schedules.

Singh (1972) compared number of pellets obtained during free (no work) and bar-pressing (work) periods for VMH lesioned, septal area lesioned, and control *S*s. Prior to surgery *S*s were randomly assigned to a FR-1, FR-3, or FR-11 schedule group and depending on the experimental condition, a lever was either present (work condition) or retracted (no-work condition) in the experimental chamber. "All rats were trained [for 10 days—5 days work; 5 days no-work] to obtain 100 reinforcements a day. However, on a given day, rats obtained reinforcement either in the work chamber or no-work chamber, but never in both . . . The rate of reinforcement of the no-work side was determined on the basis of the rat's performance on the work side [p. 260]."

Ten days after surgery *S*s received 6 additional days of training and then 4 days of preference testing (15 min/day) to test for *S*s' preference for working for pellets or receiving them without work. "If the subject selected the no-work side first, pellets were delivered at the same rate at which the subject had last worked. If the subject improved its rate of obtaining reinforcement by working during preference testing, the delivery of pellets on the no-work side was then adjusted accordingly for the next day [p. 261]." In Fig. 2 (p. 264) Singh reports the number of reinforcements (pellets) worked for by groups trained at FR-1, FR-3, or FR-11 which we have converted to mean number of presses (by multiplying number of reinforcements by schedule).

*Most of the other experiments reported here used FR schedules.

TABLE 8—ANIMAL

Activity

Source	Activity measuring instrument	Data reported in:	Normal	Dynamic	Static obese	Nature of lesion	Species	Sex	D/N	O/N
Hetherington & Ranson 1942	activity wheel	number revs	2072 (6)		160 (6)	electro-lytic	rats	male		0.08
Teitelbaum 1957	tambour mounted cage	number cage movements per day	630 (6)	275 (6)	285 (6)	electro-lytic	rats	female	0.44	0.45
Gladfelter & Brobeck 1962	running wheel	number revs	94 (12)	50 (5)		electro-lytic	rats	male	0.53	% preoperative level (during undisturbed period)
			150	135					0.90	% preoperative level (during care period)
Sclafani 1971	running wheel	number revs (×10³)	9.25 (7)	1.75 (7)		electro-lytic	rats	female	0.19	mean of 3 ten-day period
	running wheel	number revs (× 10³)	9.25 (20)	5.97 (20)		lateral knife cuts	rats	female	0.65	
Sclafani et al. 1970	open field chamber	number of squares traversed	117.5 (5)	142.5 (4)		electro-lytic	rats	female	1.22	Experiment II (mean of 2 ten-minute sessions)
			121.0 (5)	113.0 (4)					0.93	Experiment IV (mean of 5 ten-minute sessions)

DESCRIPTION OF ANIMAL STUDIES REPORTED IN TABLE 8

Activity

Hetherington and Ranson (1942) measured spontaneous running activity in a cage containing a very small living compartment and a revolving drum. All lesioned animals attaining adiposity [judged to be from + to +++ in degree of obesity in Table 1 (p. 611)] were considered static hyperphagics. Because the obesity determination was made at the end of the experiment, we only report activity data in Appendix Table 8 for the fifth to the final postoperative week after the food intake of lesioned animals appeared to reach an asymptote. No actual weight curves were provided. The mean activity data for lesioned obese and normal animals were estimated from Figures 1-4 (p. 612) from the start of the fifth postoperative week.

Teitelbaum (1957) measured random body activity in stabilimeter-type living cages for three consecutive 3-day periods—3 days of *ad lib* feeding, 3 days of food deprivation, and again 3 days of *ad lib* feeding. Data in Appendix Table 8 give the pooled mean activity per day during the two *ad lib* feeding sessions for dynamics, statics, and controls. The means are estimated from data reported in Teitelbaum's Figure 1 (p. 487).

Gladfelter and Brobeck (1962) housed animals in revolving-drum or turntable-type activity cages which recorded number of revolutions. In measuring the spontaneous locomotor activity, each week was divided into three 48-hour periods. Each period was further divided into a 43-hour period of dark and quiet and a 5-hour care period during which there was considerable activity in the laboratory. Data were recorded separately for these two periods. Each animal served as its own control and the average activity of each rat during the 43- and 5-hour periods postoperatively was expressed as a percentage of the average preoperative activity for these periods. Immediate postoperative activity (1-2 weeks after the operation) measures were not included. Data were taken from Figure 2 (p. 813) for mean activity in the undisturbed period and in the care period. Animals with medial lesions were assumed to be dynamic, since food was restricted to maintain constant body weight.

Sclafani (1971) measured locomotor activity in a running wheel pre- and post-operatively. Data were estimated from Figure 13 (p. 86) for control, VMH, and the mean scores for VL-1 and VL-2 combined.

Sclafani et al. (1970) measured locomotor activity in an open field-type chamber. (*a*) Experiment 2: Mean open field activity scores were recorded for two 10-minute sessions. Data for dynamic hyperphagics (tested two weeks after surgery) and control animals were taken from Table 1 (p. 399), pooling days 1 and 2. (*b*) Experiment 4: On the 12th postoperative day, VMH-lesioned and control rats were given five daily 10-minute tests in the open field apparatus. The data were estimated from Figure 6 (p. 401) and pooled for the 5 test days.

TABLE 8—HUMAN
Activity

Source	Activity measured	Data reported in:	Normal	Obese	Sex	O/N
Bullen et al., 1964	swimming, tennis and volleyball	% *Ss* at activity level × kilo-calories expended	3.30 (72)	1.79 (109)	female	0.54
Chirico & Stunkard 1960	walking	distance walked per day	5.96 (25)	3.74 (25)	male	0.63
			4.91 (15)	2.04 (15)	female	0.42
Johnson et al., 1956	active sports	mean hours per week	11 (28)	4 (28)	female	0.36
		mean daily index (self report)	122 (28)	96 (28)		0.79
Stefanik et al., 1959	daily activity during school year	hours per day × av. calories expended	2860 (14)	2920 (13)	male	1.02
	camp activity	% *Ss* × cals/100	68.3 (14)	69 (13)		1.01
	sports	number hours in active sports (5.5 possible)	2.9 (14)	3.2 (13)		1.10

DESCRIPTION OF HUMAN STUDIES REPORTED IN TABLE 8

Activity

Bullen et al. (1964) used motion picture photography to assess the body motion of obese and normal-weight girls at summer camp as they engaged in swimming, volley ball, and tennis. Swimming included essentially the entire camp (one camp for overweight girls and a nearby control camp for normals) on an unselected basis since specific individuals could not be recognized, while tennis and volleyball included only those girls who elected those particular sports for the day. The intensity of the motion was examined and estimated by the approximate number of kilocalories expended. The data given in Appendix Table 8 were taken from Figures 3, 4, and 5 (p. 217) and combined in the following manner: For each sport, the percentage of *S*s in each weight group at a given activity level was multiplied by the number of kilocalories represented by that level (mean levels estimated as 2.0 kcal, 2.5, 3.5, and 7, respectively). These scores were combined to provide an activity index for each sport, and indices for the three sports were pooled.

Chirico and Stunkard (1960) measured the distance walked per day by obese and normal subjects matched, in pairs, for occupation. Physical activity was measured by means of a pedometer, and the mileage walked daily was recorded for one week for female *S*s and for two weeks for male *S*s. The data reported in Appendix Table 8 were taken from Table 1 (p. 935) for obese and normal males, and from Table 2 (p. 936) for obese and normal females, and were averaged across occupation.

Johnson et al. (1956) interviewed obese and normal high-school girls to estimate average physical activity. A list of common activities was drawn up and *S*s were asked how much time they devoted to each, daily or weekly. The first row of data in Appendix Table 8 reflects the mean number of hours per week in which *S*s engaged in active sports or strenuous acts, taken from Table 2 (p. 40). The second row of data gives the mean daily activity index reported in Table 3 (p. 41). Johnson et al. derived this index by classifying activities in groups according to ratings of energy expenditure. The caloric factor was then multiplied by the total hours per week spent in each activity. The products were totaled and divided by seven to provide the daily activity index.

Stefanik et al. (1959) obtained activity records for obese and normal-weight boys for the school year and for an 8-week camp season. Twenty-four-hour activity schedules typifying school day and weekend activities were obtained by self-report at the beginning of the camp season. Activities were divided in 3 categories, each representing an approximate rate of average energy expenditure. The data in Table 6 (p. 59) give the number of hours per day *S*s engage in activities falling in these 3 categories. To obtain the activity index given on the first row of Appendix Table 8, we multiplied the hours per day at each level by the average caloric expenditure at that level (estimated at 100 Kcal, 200, and 300). These values were combined and the mean daily activity index calculated. In addition, at the end of the camp season the duration and degree of participation in two daily supervised free-choice camp activity periods were obtained from camp records submitted every week for six weeks by each boy's counselor. The time spent for activities corresponding to the three levels of exercise was expressed as a percentage of the total number of recorded periods during the 6 weeks and was reported in Table 7 (p. 59). To obtain the data reported in the second row of Appendix Table 8 these percentages divided by 100 were multiplied by the average caloric expenditure at each activity level. Scores were then combined and a mean daily activity estimate was calculated. The final row of Appendix Table 8 represents the average number of hours (out of a possible 5.5 per day) that each weight group engaged in active sports. These data were reported in the text on p. 60.

TABLE 9—ANIMAL

Emotionality

Source	Measure	Data reported in:	Normal	Dynamic	Static obese	Lesion	Species	Sex	D/N	O/N
Grossman 1972	response to foot shock of various intensities	probability of fighting in response to foot shocks	0.14 (10)		0.61 (20)	electro-lytic	rats	female		4.36
Sclafani 1971	rated irrita-bility and re-activity to handling	reactivity scores	0 (12)	7.7 (13)	5.5 (13)	electro-lytic	rats	female		
			0 (12)	4.33 (21)	1.62 (21)	ventro-lateral knife cuts	rats	female		
Singh 1969	ratings of response to handling, capturing, tap on back, vocalization, presentation with pencil	level of emotionality on 6 point rating scale	0.30 (12)	1.62 (12)	2.04 (12)	electro-lytic	rats	female	5.40	6.80

TABLE 9—ANIMAL (Continued)

Source	Measure	Data reported in:	Normal	Dynamic	Static obese	Lesion	Species	Sex	D/N	O/N
Eichelman 1971	attack in response to different intensities of shock	difference in % of attacks before and after lesion	4.7 (20)	22.1 (20)		electro-lytic	rats	male	4.7	
	jump-flinch thresholds	mamps	.26 (20)	.17 (20)		electro-lytic	rats	male	.65	
Paxinos & Bindra 1972	biting gloved hand within 20 sec.	% of biting/ opportunities for biting	0 (10)	.91 (10)		para-sagittal knife cuts separating medial from lateral hypothalamic areas	rats	male		
Turner et al., 1967	flinch threshold	mamps	.129 (11)	.075 (11)		GTG	mice	male	.58	
	jump, jerk, vocalize threshold	mamps	.208	.147					.71	
	flinch threshold	mamps	.56 (10)	.32 (10)		electro-lytic	rats	male	.57	
	jump, jerk vocalize threshold	mamps	1.00	.56					.56	

DESCRIPTION OF ANIMAL STUDIES REPORTED IN TABLE 9

Emotionality

Grossman (1972) measured emotionality by scoring response to shock in paired VMH animals and paired controls. A "threat" was scored whenever both rats assumed an upright posture and faced each other. An "attack" was recorded when one rat hit, struck, or bit the other. Both classes of aggressive responses were combined to score "fighting" (p. 275). Pairs of control rats and pairs of lesioned rats were tested 6 weeks after surgery and so VMH animals are tentatively considered static. Six 10-trial tests were given with 5-min. rest periods between alternating ascending or descending series of 10 shocks (.01, .02, .05, .18, .23, .32, .35, .38, and .40 ma.). Mean probability of fighting scores were computed from data in Figure 2. (p. 276)

Sclafani (1971) rated the emotionality of VMH lesioned, VL-1 and VL-2 parasagittal-cut *S*s (see Sclafani, 1971, in Appendix Table 3 for exact placement of these transections) and sham operated *S*s, as reactivity to handling. "Beginning with a first postoperative day and continuing for 8 weeks, the subjects' reactivity to handling was scored during the daily care period using a scale containing six categories: 1) resistance to capture, 2) resistance to handling, 3) muscular tension reaction to capture and handling, 4) vocalization, 5) urination and/or defecation, and 6) aggressive reaction to prodding of snout with a rod. Scores of from 0 (no reactivity) to 3 (maximum reactivity) were given for each of the six categories with a possible daily score ranging from 0-18 [p. 73]." The data in Appendix Table 9 are divided into the period when the hyperphagics were considered dynamic (1-6 weeks post operation) and static (7-8 weeks post operation), judging by estimates from the weight curves reported in Figure 7 (p. 77). Data are taken from Figure 8 (p. 80) and pooled for the VL-1 and VL-2 *S*s.

Singh (1969) rated emotionality and ventromedial lesioned animals, using a six-point rating scale for emotional reactivity. All *S*s were rated by two independent observers on five indices of emotionality: resistance to handling, resistance to capture, startle and flight reaction to a light tapping on the animal's back with a lead pencil, vocalization reaction when captured and handled, and attack or flight reaction to a lead pencil held directly in front of the animal. Animals were measured pre- and postoperatively. The data in Appendix Table 9 were estimated from Figure 1 (p. 3) for the handled ventromedial group. Data for normals were taken from the preoperative measures. The data for dynamic hyperphagics were obtained from tests given on postoperative days 1-8. A single measure of emotionality on day 45 provided the static hyperphagic data. Data for the nonhandled group were not included in our calculations.

Eichelman (1971)

(*a*) Experiment I: Eichelman measured emotionality as a change in shock-induced aggression, expressed as number of attacks, following VMH lesioning. "Each daily session consisted of delivering 50 footshocks to each pair of rats and counting the number of attack responses made . . . Aggression attacks were defined as directed movement toward the opponent which resulted in contact, including at least one additional response of the following: biting, sparring upright attack posture, or supine submissive posturing, adopted by the attacked rat [p. 333]." There were ten daily sessions of fighting pre- and postoperation and mean percentages (attack/shock × 100) were calculated for each ten-day period. Although electrolytic lesions were given in 8 different locations, we report data from the VMH and operated control groups only. Data for attack scores are taken from Table 1 (p. 335). Testing took place 3-8 days after surgery when the VMH rats were dynamic.

(*b*) Experiment II: *S*s from Experiment 1 were also tested in this experiment. After all other testing was completed and the animals were still in the dynamic stage of obesity, jump-flinch measurements were taken. Each rat was given 10 series of unavoidable shock of six intensities ranging from .07-.5 ma. in alternating ascending-descending order. Each shock was 4 sec. in duration, with a 30 sec. interval between shocks and a 2 min. interval between series. The *S*s response was recorded as 0 for no response, 1 for flinch, and 2 for jump. Flinch and jump thresholds were calculated for each animal at the lowest shock level which elicited the response in 50% of the series. Data reported in Appendix Table 9 are taken from Table 2 (p. 336).

Paxinos and Bindra (1972) measured irritability by scoring 0 for no biting of a gloved hand pushing the animal back and 1 for biting the gloved hand within 20 sec. Experimental *S*s received bilateral parasagittal knife cuts separating the medial from the lateral hypothalamic areas. The post-surgery irritability scores were taken 3 weeks after surgery when the rates were still gaining weight (Fig. 2, p. 222) and in the dynamic stage of hyperphagia. Data reported in Appendix Table 9 are taken from p. 223 in the text of the article and are reported as a percentage (bites actually given out of the total number of opportunities for biting).

Turner et al. (1967) tested goldthioglucose-lesioned mice and electrolytically lesioned rats. Lesioned and control animals were given four sets of threshold tests with each set given 24 hours apart. A set consisted of shock intensities ranging from 0.05 to 0.25 mamp for the mice and 0.1 to 1.5 for the rats. The shock intensities of the first set were in ascending order, the second in descending order, and the last two, randomized. Following each shock a response was recorded as: no response; flinch; or jerk, jump, or vocalize. An average of the values from the four sets provided the threshold level for each animal. Data for flinch, and data for jerk, jump, or vocalize thresholds were reported in Tables 1 and 2 (p. 239 and 240). No weight data were given; however, the authors did report an average of 1 month elapsed between GTG administration and testing and about 2 weeks between electrolytic lesioning and shock threshold determination.

TABLE 10—ANIMAL
Active Avoidance

Source	Apparatus	Data reported in:	Normal	Dynamic	Static obese	Nature of lesion	Species	Sex	D/N	O/N
Grossman 1966	shuttlebox	% avoidance over all trials	42.1 (8)	75.1 (12)		atropine	rats	male	1.78	
Grossman 1972	shuttlebox	# trials to meet increasing-ly stringent criteria	73.11 (10)		26.22 (20)	electro-lytic	rats	female		2.79 [a]
Grossman 1970	shuttlebox	# trials to meet increasing-ly stringent criteria	54.56 (6)	143.67 (6)		knife cuts	rats	female	0.38[a]	
Levine & Soliday 1960	shuttlebox	% correct avoidance responses to 12 five-trial blocks	22.6 (11)	48.5 (8)		electro-lytic	rats	male	2.15	
McAdam & Kaelber 1966	shuttlebox	# anticipatory jumps per block	5.42 (15)	3.6 (9)		electro-lytic	cats	?	0.66	
Sechzer et al., 1966	electrified Y-maze	trials to 80% or greater cor-rect responses for 3 consec. days	145 (20)	70 (19)		GTG	mice	male	2.07[a]	
Sepinwall 1969	noncued (Sidman) avoidance apparatus	responses per min.	27.7 (10)	66.5 (10)		atropine	rats	male	2.4	

[a]To maintain consistency with other ratios, where a figure greater than 1 indicates superiority in the lesioned animal, this ratio was calculated for N/O or N/D in those studies where the data were reported in number of trials to criterion.

DESCRIPTION OF ANIMAL STUDIES REPORTED IN TABLE 10

Active Avoidance

Grossman (1966) used a shuttle box apparatus to train avoidance responses in atropine-treated and control animals. During training, the CS (an overhead light) was presented for 5 seconds followed by a 55-second CS-UCS (shock) period. The subjective intertrial interval was determined by the response latency of each animal but could not exceed 60 seconds. The shock level, judged to be only mildly noxious on the basis of preliminary testing, was set at 250 microamps. Two hundred and forty acquisition trials (15 massed trials/day for 16 days) were used. Experimental animals (nonobese) received atropine 5 minutes before the first trial each day. Data were estimated from Figure 4 (p. 8) which reports the % avoidance responses over the 240 acquisition trials in 30 trial blocks.

Grossman (1972) trained VMH electrolytic-lesioned rats in a shuttlebox avoidance task with a light CS and a UCS of 275 microamps of foot shock which was pulsed at a rate of 1/sec. with an "on" time of ½ sec. At 6-10 weeks after surgery *S*s were trained using 15 CS-UCS presentations per day (with an intertrial interval of 60 sec) for 14 consecutive days "or until each animal had reached a criterion of nine avoidance responses on 10 consecutive trials on each of 2 consecutive days [p. 279]." Appendix Table 10 reports number of trials required to meet increasingly stringent criteria of avoidance proficiency, estimated from Figure 3 (p. 279). Animals were considered static since testing occurred 6-10 weeks after surgery.

Grossman (1970) tested adult female rats given parasagittal surgical cuts (isolating the medial half of the hypothalamus from all lateral connections) in a shuttlebox avoidance task. Six weeks after surgery, *S*s were trained for 21 days with 15 trials per day. The CS was a 7W light and the UCS consisted of pulses (0.3 sec. on, 0.3 sec. off) of constant current shock. Shock followed 5 sec. after the light went on and the CS-UCS combination remained on for 58 seconds before switching to the previously safe dark compartment. Subjects had been aphagic and adipsic for 2-8 days after surgery and hyperphagic only on a preferred pellet diet. At the time of the avoidance training they were maintaining normal body weight and are loosely designated as dynamic. The data are taken from Figure 3 (p. 1106) and reported in mean number of trials required to meet increasingly stringent criteria of avoidance proficiency (such as 1, 2, 3, ... 9 avoidance responses on 10 consecutive trials.

Levine and Soliday (1960) tested rats with VMH electrolytic lesions and unoperated and operated controls in a conditioned avoidance learning task. A shuttlebox was used for testing with a small buzzer as the CS and an "unlimited shock" of 3 ma. as the UCS. ". . . . all *S*s were run in blocks of five trials, with an interval of 30 min. between blocks. The CS-UCS interval was 5 sec., and the intertrial interval was randomly varied from 15 to 30 sec. . . Six blocks of trials (30 trials) were given to each *S* on two successive days (60 trials). The criterion of learning was set at 9 out of 10 avoidance responses with the last 5 in the run correct [p. 498]." Data in Appendix Table 10 compare nonoperated controls to VMH. *S*s are taken from Figure 3 (p. 499). Since testing was conducted two days after surgery, experimental *S*s are considered dynamic hyperphagics.

McAdam and Kaelber (1966) trained lesioned (2 weeks postoperatively) and control cats in a shuttle box avoidance situation, consisting of 24 trials per day for 3 consecutive days. The CS, a tone, lasted for 20 secs. followed by a 60 sec. CS-UCS interval. The intertrial interval ranged from 30 to 90 secs. Data from Figure 2 (p. 296) represent the average of the means for all 8 trial blocks. Data for all experimental animals who learned (unfriendly and savage) were combined.

Sechzer et al. (1966) trained goldthioglucose-lesioned mice (22 days post operatively) and controls in an electrified Y-maze. In order to avoid shock, the *S* had to move from the start box, down the alley, and into the correct arm of the maze within 5 seconds. Each animal was trained at its own shock threshold level. (Thresholds were lower for lesioned animals than for controls.) Twenty training trials a day were given until a criterion of 80% or more correct responses was reached and maintained for 3 consecutive days. Data were reported in mean number of trials to reach criterion and are estimated from Figure 2 (p. 260).

TABLE 11—ANIMAL

Passive Avoidance

Source	Apparatus	Data reported in:	Normal	Dynamic	Static obese	Nature of lesion	Species	Sex	N/D[a]
Kaada et al., 1962	shock delivered through electrified grid floor	number shocked responses	2.4 (24)	37.0 (5)		electro-lytic	rats	male & female	0.06
Margules & Stein 1969	Foringer response chamber; shock de-livered through electrified grid floor	number shocked responses	7.3 (3)	20.4 (3)		atropine	rats	male	0.36
Sclafani & Grossman 1971	shock delivered through drinking tube in test chamber	number shocked responses	9 (7) 6	13 (7) 23		electro-lytic	rats	female	0.69 PA 1 Exper. I 0.26 PA 2
		number shocked responses	4 (8) 5	8 (8) 12					0.50 PA 1 & 3 Exper. II 0.42 PA 2
		amount drunk in ml	3.7 (10)	5.7 (10)					0.65 Exper. III

[a] A ratio greater than 1 has previously indicated superiority for the lesioned animal. To keep this consistent, all the ratios are given here for N/D.

Sepinwall (1969) tested the effects of various drugs, applied directly through cannula implants, on performance in a modified Sidman avoidance task. Pre- and post-drug response rates were reported for atropine application to the VMH in Table 1 (p. 395) and these were taken to reflect normal and dynamic response rates, respectively. Shocks, presented through a grid floor, were .3-.4 ma. in intensity and 90 msec. in duration and were repeated once a second (S-S interval of .9 sec.). A response, the turn of a wheel, postponed the next shock for 1 sec. If there were 4 or more responses within 1 sec., shock was delayed for 20 secs. Since relatively stable performance occurs rapidly with this schedule, data reflect performance rather than acquisition rates.

DESCRIPTION OF ANIMAL STUDIES REPORTED IN TABLE 11

Passive Avoidance

Kaada et al. (1962) tested lesioned (6-8 weeks postoperatively) and control animals in an apparatus which provided a .183 ma. (475 v., 50 cps., with an external resistance of 2.6 megohms) current whenever *S*s drank water from a dish in a corner of the cage. During testing, animals were fed dry rat pellets and no water. After 48 hours of water deprivation animals were placed in the test apparatus with the electrified water dish for 20 minutes. The mean number of shocks received by animals with hypothalamic lesions was taken from data reported in the text on p. 663. Three of these 5 animals had lesions confined only to the ventromedial nuclei and their immediate surroundings. Since it was impossible to separate the data for these 3 animals from the other 2, we report the data for all 5.

Margules and Stein (1969) trained animals in a passive avoidance situation which alternated four 18-minute punished and unpunished reinforcement schedules. In the unpunished schedule, a lever press response was rewarded with 22 cc of milk at VI-2 min. In the punished schedule (signaled by a continuous tone), every response was rewarded with milk and also punished by a .25 sec. shock to the paws. The shock intensity was increased gradually until responses in the tone period were almost completely suppressed. Cannulae were implanted after performance had stabilized for at least 2 weeks. Experimental animals received atropine 15 and 1 minutes before testing. The same animals provided control data when they were sham stimulated. The data for experimental *S*s were placed in the dynamic column since they were not obese during testing, but it is not clear whether they ever overate and gained weight with the atropine preparation. Data for mean number of shocked responses were reported in the text on p. 330.

Sclafani and Grossman (1971) tested lesioned and control animals in a variety of passive avoidance procedures:

(*a*) Experiment 1: In PA test 1, animals with *ad lib* food and water were given nine daily 20-minute test periods using an 8% dextrose-water solution. On the final 2 days shock was delivered through the drinking tube. In PA test 2, which followed PA test 1, all *S*s were placed on a 22.6-hour water deprivation schedule. For 10 days animals were given access to tap water during a 20-minute test period and during the last 4 days, shock was delivered through the drinking tube. Both tests used .2 mamp current. The data were estimated from Figure 2 (p. 160).

(*b*) Experiment 2: PA 1 and 3 are pooled for the present analysis since in both tests animals were placed on a 22.6-hour water deprivation schedule and a restricted food diet, adjusted so that control and hyperphagic animals consumed equivalent amounts of water. In PA 1, animals were given 11 daily test sessions and on the last three, shock was also delivered through the drinking tube. In PA 3 there were 5 nonshock, followed by 3 shock, days.

In PA 2, animals were placed on a 22.6-hour water deprivation-*ad lib* food schedule. In all cases they were given free access to tap water during the 20-minute test sessions. Data were estimated from Figure 4 (p. 161). The shock intensity was .2 mamp.

(*c*) Experiment 3: Animals were water deprived and were allowed access to tap water during 30-minute test sessions occurring every 12 hours. After 5 or 6 nonshock sessions, test periods began using different, alternating shock intensities (.035, .010, and .025). The data were reported on p. 162.

TABLE 12—ANIMAL
The Effects of Caloric Dilution on Intake

Source	Test food	Data reported in:	Amount eaten by: Normal	Dynamic	Static obese	Nature of lesion	Species	Sex	D/N	O/N	Dilution
Solid Food											
Carlisle & Stellar 1969	pellets	gms	26 (4)		30 (4)	electro- lytic	rats	male		1.16	no dilution
	powdered diet with crisco oil	gms	21.5 19.3		24 28	electro- lytic				1.12 1.45	5+15% 25+35+45%
	powdered diet with isocaloric oil	gms	21 25.3		20 35.3	electro- lytic				0.95 1.40	5+15% 25+35+45%
	powdered diet with mineral oil	gms	22 36.3		20.5 43	electro- lytic				0.93 1.19	5+15% 25+35+45%
Teitelbaum 1955	regular powdered diet	gms	12 (6)	31 (6)	21 (6)	electro- lytic	rats	female	2.58	1.75	no dilution
	powdered diet with cellulose		14.7 9.5	23.3 13.5	15 1				1.59 1.42	1.02 0.11	5+15+25% 50+75%
Singh & Meyer 1968	regular powdered diet	pre-minus post-lesion intake in grams	0.8 (8)	8.45 (8)		electro- lytic	rats	male & female	10.5		no dilution
	powdered diet with cellulose		5.48	3.23					0.59		12.5+25%

TABLE 12—ANIMAL (Continued)

Source	Test food	Data reported in:	Amount eaten by: Normal	Dynamic	Static obese	Nature of lesion	Species	Sex	D/N	O/N	Dilution
Smith et al., 1961	pellets	gms	0.99 (9)	0.86 (8)	0	electro-lytic	rats	female	0.87		glucose 1 molar/ 10 cc (after 10 & 30 minutes)
			7.32 (9)	5.00 (8)					0.68		(after 120 minutes)
Liquid food											
Thomas & Mayer 1968	liquid diet	ml	72 (6)	124 (6)		atropine	rats	female	1.73		Experiment II no dilution
	liquid diet diluted with water		62	107					1.72		
	liquid diet	ml	63 (6)	114 (6)					1.80		Experiment III no dilution
	liquid diet diluted with water		49	67					1.38		
Williams & Teitelbaum 1959	liquid diet	ml	45.7 (6)	78.0 (3)		electro-lytic	rats	female	1.71		no dilution
	liquid diet diluted with 50% water		75.8	140.5					1.85		50% water

Lesioned animals were considered dynamic hyperphagics because of the short postoperative period and the rigorous deprivation regime. They were, however, significantly heavier than controls.

DESCRIPTION OF ANIMAL STUDIES DESCRIBED IN TABLE 12

The Effects of Caloric Dilution on Intake

Carlisle and Stellar (1969) diluted powdered food with Crisco oil (increased caloric density as oiliness increased), mineral oil (decreased caloric density as oiliness increased), and isocaloric oil (maintained constant caloric density as oiliness increased). An *ad lib* baseline for intake on pellets was provided for comparison. The data were estimated from Figure 1 (p. 109). We pooled data for the 5% and 15% oil dilutions and compared them to the combined means for 25%, 35%, and 45% dilutions. Each type of oil dilution was considered separately. It seems apparent that increasing liquidity may have covaried with increasing caloric dilution.

Teitelbaum (1955) mixed standard powdered diet with varying amounts of nonnutritive cellulose (0, 5, 15, 25, 50, and 75%) and presented the food *ad lib*. The data were estimated from Figure 2 (p. 158). In Appendix Table 12, a baseline level of intake is given in the first row. In the second row, pooled data for 5%, 15%, and 25% dilutions are given, and in the third row, we report the average for 50% and 75% dilutions.

Singh and Meyer (1968) presented powdered food adulterated with 0%, 12.5%, or 25% of powdered cellulose during the 17-day test period after ventromedial hypothalamic lesions were made in experimental animals. During 9 days of this period, each of the three dilutions was given on 3 different days, in counterbalanced order. The data were estimated from Figure 2 (p. 165) for low (0%) baseline vs. medium (12.5%) and high (25%) cellulose dilutions combined. Data for male and female animals were pooled.

Smith et al. (1961) measured grams of pellets eaten by control and lesioned animals at 10, 30, and 120 minutes after *S*s received liquid preloads through a stomach tube. Testing occurred approximately 3 weeks after surgery, and after the weights of the hyperphagic animals were reduced and maintained at levels equal to control *S*s. These dynamic hyperphagic *S*s and the control *S*s were put on a 2-hour/day eating schedule and their responses to the various preloads were measured. Three cycles of tests were run on the same group of *S*s, the three conditions (preloads of 10cc. in each case) were water, a one-molar solution of glucose, and a one-half molar solution of sodium chloride presented in counterbalanced order for each *S*. Data for the glucose preload conditions presented in Appendix Table 12 are taken from Table 1 (p. 662). Measurements of amounts eaten at 10 and 30 minutes after preload were pooled.

Thomas and Mayer (1968) gave liquid diet, delivered when the *S* pressed a bar, to dynamic and control animals. During testing, the liquid diet was infused through a gastric tube such that 50-75% of *S*'s caloric intake prior to infusion was provided intragastrically. Data for intake volume in ml. were taken from Table 1 (p. 647). It was estimated from data reported in the procedure section (p. 647) that normals were receiving on the average 22.5 ml. intragastrically and hyperphagics were getting 40 ml., pooling 50 and 75% IG conditions.

In a second experiment, oral consumption of a diluted diet was paired with gastric infusion of a more concentrated solution. Each bar press delivered 1 cal/ml. orally and 2 cal/ml. IG. Data were taken from Table 2 (p. 649)

Williams and Teitelbaum (1959) diluted liquid diet with 50% water and measured the intake of normals and dynamic hyperphagics. Data in Appendix Table 12 are estimated from Figure 1 (p. 460) pooling the means for intake on the full strength diet and pooling those on days 4-7 when the half strength diet was given. Food was available continuously and was delivered through glass drinking tubes. Water was always available.

REFERENCES

Anand, B.K., & Brobeck, J.R. Hypothalamic control of food intake in rats and cats. *Yale Journal of Biology and Medicine,* 1951, **24,** 123-140.

August, R.V. *Hypnosis in obstetrics.* New York: McGraw-Hill, 1961.

Balagura, S., & Devenport, L.D. Feeding patterns of normal and ventromedial hypothalamic lesioned male and female rats. *Journal of Comparative and Physiological Psychology,* 1970, **71,** 357-364.

Barber, T.X., & Hahn, K.W. Physiological and subjective responses to pain producing stimulation under hypnotically-suggested and waking-imagined analgesia. *Journal of Abnormal and Social Psychology,* 1962, **65,** 411-418.

Beaudoin, R., & Mayer, J. Food intakes of obese and non-obese women. *Journal of the American Dietetic Association,* 1953, **29,** 29-33.

Brobeck, J.R. Mechanisms of the development of obesity in animals with hypothalamic lesions. *Physiological Review* 1946, **26,** 541-559.

Brobeck, J.R., Tepperman, J., & Long, C.N.H. Experimental hypothalamic hyperphagia in the albino rat. *Yale Journal of Biology and Medicine,* 1943, **15,** 831-853.

Brooks, C.McC., & Lambert, E.F. A study of the effect of limitation of food intake and the method of feeding on the rate of weight gain during hypothalamic obesity in the albino rat. *American Journal of Physiology,* 1946, **147,** 695-707.

Brooks, C.McC., Lockwood, R.L., & Wiggins, M.L. A study of the effect of hypothalamic lesions on the eating habits of the albino rat. *American Journal of Physiology,* 1946, **147,** 735-741.

Brown, J.D., & Pulsifier, D.H. Outpatient starvation in normal and obese subjects. *Aerospace Medicine,* March, 1965, 267-269.

Bullen, B.A., Reed, R.B., & Mayer, J. Physical activity of obese and nonobese adolescent girls appraised by motion picture sampling. *American Journal of Clinical Nutrition,* 1964, **14,** 211-223.

Carlisle, H., & Stellar, E. Caloric regulation and food preference in normal, hyperphagic, and aphagic rats. *Journal of Comparative and Physiological Psychology,* 1969, **69,** 107-114.

Chirico, A., & Stunkard, A. Physical activity and human obesity. *New England Journal of Medicine,* 1960, **263,** 935-940.

Corbit, J. Hyperphagic hyperreactivity to adulteration of drinking water with quinine HCL. *Journal of Comparative and Physiological Psychology,* 1965, **60,** 123-124.

Corbit, J., & Stellar, E. Palatability, food intake, and obesity in normal and hyperphagic rats. *Journal of Comparative and Physiological Psychology,* 1964, **58,** 63-67.

Cox, V.C., Kakolewski, J.W., & Valenstein, E.S. Ventromedial hypothalamic lesions and changes in body weight and food consumption in male and female rats. *Journal of Comparative and Physiological Psychology,* 1969, **67,** 320-326.

Decke, E. Effects of taste on the eating behavior of obese and normal persons. Cited in S. Schachter, *Emotion, obesity, and crime.* New York: Academic Press, 1971, p. 103.

Duncan, G., Jinson, W., Fraser, R., & Christori, F. Correction and control of intractable obesity. *Journal of the American Medical Association*, 1962, **181**, 309-312.

Eichelman, B. Effect of subcortical lesions on shock-induced agression in the rat. *Journal of Comparative and Physiological Psychology*, 1971, **74**, 331-339.

Falk, J.L. Comments on Dr. Teitelbaum's paper. *Nebraska symposium on motivation*, 1961, **9**, 65-68.

Ferguson, N.B.L., & Keesey, R.E. Comparison of ventromedial hypothalamic lesion effects upon feeding and lateral hypothalamic self stimulation in the female rat. *Journal of Comparative and Physiological Psychology*, 1971, **74**, 263-271.

Gladfelter, W.E., & Brobeck, J.R. Decreased spontaneous locomotor activity in the rat induced by hypothalamic lesions. *American Journal of Physiology*, 1962, **203**, 811-817.

Gold, R.M. Hypothalamic hyperphagia: Males get just as fat as females. *Journal of Comparative and Physiological Psychology*, 1970, **71**, 347-356.

Goldman, R.L. The effects of the manipulation of the visibility of food on the eating behavior of obese and normal subjects. Unpublished doctoral dissertation, Columbia University, 1968.

Goldman, R., Jaffa, M., & Schachter, S. Yom Kippur, Air France, dormitory food, and the eating behavior of obese and normal persons. *Journal of Personality and Social Psychology*, 1968, **10**, 117-123.

Graff, H., & Stellar, E. Hyperphagia, obesity, and finickiness. *Journal of Comparative and Physiological Psychology*, 1962, **55**, 418-424.

Greene, R.J., & Rehner, J. Pain tolerance in hypnotic analgesic and imagination states. *Journal of Abnormal Psychology*, 1972, **79**, 29-38.

Grossman, S.P. The VMH: A center for affective reactions, satiety, or both? *Journal of Physiology and Behavior*, 1966, **1**, 1-10.

Grossman, S.P., *A textbook of physiological psychology*, New York: Wiley, 1967.

Grossman, S.P. Avoidance behavior and aggression in rats with transections of the lateral connections of the medial or lateral hypothalamus. *Physiology and Behavior*, 1970, **5**, 1103-1108.

Grossman, S.P. Aggression, avoidance and reaction to novel environments in female rats with ventromedial hypothalamic lesions. *Journal of Comparative and Physiological Psychology*, 1972, **78**, 274-283.

Hamilton, C.L., & Brobeck, J.R. Hypothalamic hyperphagia in the monkey. *Journal of Comparative and Physiological Psychology*, 1964, **57**, 271-278.

Han, P.W., Lin, C.H., Chu, K.C., Mu, J.Y., & Liu, A.C. Hypothalamic obesity in weanling rats. *American Journal of Physiology*, 1965, **209**, 627-631.

Hetherington, A.W., & Ranson, S.W. Hypothalamic lesions and adiposity in the rat. *Anatomical Record*, 1940, **78**, 149-172.

Hetherington, A.W., & Ranson, S.W. The spontaneous activity and food intake of rats with hypothalamic lesions. *American Journal of Physiology*, 1942, **136**, 609-617.

Johnson, M.L., Burke, B.S., & Mayer, J. Relative importance of inactivity and overeating in the energy balance of obese high school girls. *American Journal of Clinical Nutrition*, 1956, **4**, 37-44.

Kaada, B.R., Rasmussen, E.W., & Kveim, O. Impaired acquisition of passive avoidance behavior by subcallosal, septal, hypothalamic, and insular lesions in rats. *Journal of Comparative and Physiological Psychology*, 1962, **55**, 661-670.

Karp, S.A., & Pardes, H. Psychological differentiation (field dependence) in obese women. *Psychosomatic Medicine*, 1965, **27**, 238-244.

Kennedy, G.C. The hypothalamic control of food intake in rats. *Proceedings, Royal Society* (London), B, 1950, **137**, 535-548.

Kling, A., & Hutt, P.J. Effect of hypothalamic lesions on the amygdala syndrome in the cat. *Archives of Neurology and Psychiatry*, 1958, **79**, 511-517.

Larkin, R.P., & Kissileff, H.R. Hyperphagic rats: Will they work for food? Unpublished manuscript, University of Pennsylvania, 1971.

Larsson, V., & Strom, L. Some characteristics of goldthioglucose obesity in the mouse. *Acta Physiologica Scandinavia*, 1957, **38**, 298-308.

Levine, S., & Soliday, S. The effects of hypothalamic lesions on conditioned avoidance learning. *Journal of Comparative and Physiological Psychology*, 1960, **53**, 497-501.

Lipton, J.M. Effects of high fat diets on caloric intake, body weight, and heat escape responses for normal and hyperphagic rats. *Journal of Comparative and Physiological Psychology*, 1969, **68**, 507-515.

Lykken, D. A study of anxiety in the sociopathic personality. *Journal of Abnormal and Social Psychology*, 1957, **55**, 6-10.

MacLean, P.D. The hypothalamus and emotional behavior. In W. Haymaker, E. Anderson, & W.J.H. Nauta (Eds.), *The hypothalamus*. Springfield, Ill.: C.C. Thomas, 1969.

Maller, O. The effect of hypothalamic and dietary obesity on taste preference in rats. *Life Sciences*, 1964, **3**, 1281-1291.

Margules, D.L., & Stein, L. Cholinergic synapses in the ventromedial hypothalamus for the suppression of operant behavior by punishment and satiety. *Journal of Comparative and Physiological Psychology*, 1969, **68**, 327-335.

Marks, H.E., & Remley, N.R. The effects of type of lesion and percentage body weight loss on measures of motivated behavior in rats with hypothalamic lesions. *Behavioral Biology*, 1972, **7**, 95-111.

McAdam, D.W., & Kaelber, W.W. Differential impairment of avoidance learning in cats with ventromedial hypothalamic lesions. *Experimental Neurology*, 1966, **15**, 293-298.

Metropolitan Life Insurance Company. New weight standards for men and women. *Statistical Bulletin*, 1959, **40**, 1-4.

Miller, N.E. Some psychophysiological studies of motivation and of the behavioural effects of illness. *Bulletin of the British Psychological Society*, 1964, **17**, 1-20.

Miller, N.E., Bailey, C.J., & Stevenson, J.A.F. Decreased "hunger" but increased food intake resulting from hypothalamic lesions. *Science*, 1950, **112**, 256-259.

Moore, M.E., Stunkard, A., & Srole, L. Obesity, social class & mental illness. *Journal of the American Medical Association*, 1962, **181**, 962-966.

Mrosovsky, N. *Hibernation and the hypothalamus*. New York: Appleton-Century-Crofts, 1971.

Nachman, M. Hypothalamic hyperphagia, finickiness, and taste preferences in rats. *Proceedings of the 75th Annual Convention of the American Psychological Association*, 1967, **2**, 127-128.

Nisbett, R.E. Determinants of food intake in human obesity. *Science*, 1968, **159**, 1254-1255. (a)

Nisbett, R.E. Taste, deprivation, and weight determinants of eating behavior. *Journal of Personality and Social Psychology*, 1968, **10**, 107-116. (b)

Nisbett, R.E. Eating and obesity in men and animals. In press.

Nisbett, R.E., & Gurwitz, S. Weight, sex, and the eating behavior of human newborns. *Journal of Comparative and Physiological Psychology*, 1970, **73**, 245-253.

Nisbett, R.E., & Schachter, S. The cognitive manipulation of pain. *Journal of Experimental Social Psychology*, 1966, **2**, 227-237.

Nisbett, R.E., & Storms, M.D. Cognitive, social, and physiological determinants of food intake. In H. London & R.E. Nisbett (Eds.), *Cognitive modification of emotional behavior*. Chicago: Aldine, in press.

Ornstein, R. *On the experience of time*. Middlesex, England: Penguin Books Ltd., 1969.

Palka, Y., Liebelt, R.A., & Critchlow, V. Obesity and increased growth following partial or complete isolation of ventromedial hypothalamus. *Physiology and Behavior*, 1971, **7**, 187-194.

Paxinos, G., & Bindra, D. Hypothalamic knife cuts: Effects on eating, drinking, irritability, aggression, and copulation in the male rat. *Journal of Comparative and Physiological Psychology*, 1972, **79**, 219-229.

Penick, S.B., & Stunkard, A. Newer concepts of obesity. *Medical Clinics of North America*, 1970, **54**, 745-754.

Pliner, P. The effects of cue salience on the behavior of obese and normal subjects. *Journal of Abnormal Psychology*, 1973, **82**, 226-232. (a)

Pliner, P. Effect of external cues on the thinking behavior of obese and normal subjects. *Journal of Abnormal Psychology*, 1973, **82**, 233-238. (b)

Porter, J., & Allen, J. Food motivated performance as a function of weight loss in hypothalamic hyperphagic rats. *Psychonomic Science*, 1972, **28**, 285-288.

Rabin, B.M., & Smith, C.J. Behavioral comparison of the effectiveness of irrative and non-irritative lesions in proudcing hypothalamic hyperphagia. *Physiology and Behavior*, 1968, **3**, 417-420.

Reynolds, R.W. Ventromedial hypothalamic lesions without hyperphagia. *American Journal of Physiology*, 1963, **204**, 60-62.

Roazin, I.H. Effects on pain threshold of concomitant distraction and past distraction. Unpublished doctoral dissertation, University of Minnesota, 1967.

Ross, L.D. Cue- and cognition-controlled eating among obese and normal subjects. Unpublished doctoral dissertation, Columbia University, New York, 1969.

Ross, L.D., Pliner, P., Nesbitt, P., & Schachter, S. Patterns of externality and internality in the eating behavior of obese and normal college students. Unpublished manuscript, Columbia University. Cited in S. Schachter, *Emotion, obesity, and crime,* New York: Academic Press, 1971.

Ross, L.D., Rodin, J., & Zimbardo, P. Toward an attribution therapy: The reduction of fear through induced cognitive-emotional misattribution. *Journal of Personality and Social Psychology,* 1969, **12,** 279-288.

Schachter, S. Obesity and eating. *Science,* 1968, **161,** 751-756.

Schachter, S., *Emotion, obesity, and crime.* New York: Academic Press, 1971. (a)

Schachter, S. Some extraordinary facts about obese humans and rats. *American Psychologist,* 1971, **26,** 129-144. (b)

Schachter, S., Goldman, R., & Gordon, A. Effects of fear, food deprivation, and obesity on eating. *Journal of Personality and Social Psychology,* 1968, **10,** 91-97.

Schachter, S., & Gross, L. Manipulated time and eating behavior. *Journal of Personality and Social Psychology,* 1968, **10,** 98-106.

Sclafani, A. Neural pathways involved in the ventromedial hypothalamic lesion syndrome in the rat. *Journal of Comparative and Physiological Psychology,* 1971, **77,** 70-96.

Sclafani, A., Belluzzi, J.D., & Grossman, S.P. Effects of lesions in the hypothalamus and amygdala on feeding behavior in the rat. *Journal of Comparative and Physiological Psychology,* 1970, **72,** 394-403.

Sclafani, A., & Grossman, S.P. Hyperphagia produced by knife cuts between the medial and the lateral hypothalamus in the rat. *Physiology and Behavior,* 1969, **4,** 533-537.

Sclafani, A., & Grossman, S.P. Reactivity of hyperphagic and normal rats to quinine and electric shock, *Journal of Comparative and Physiological Psychology,* 1971, **74,** 157-166.

Sechzer, J.A., Turner, S.G., & Liebelt, R.A. Motivation and learning in mice after goldthioglucose-induced hypothalamic lesions. *Psychonomic Science,* 1966, **4,** 259-260.

Sepinwall, J. Enhancement and impairment of avoidance behavior by chemical stimulation of the hypothalamus. *Journal of Comparative and Physiological Psychology,* 1969, **68,** 393-399.

Siegel, S. *Nonparametric statistics for the behavioral sciences.* New York: McGraw-Hill, 1956.

Singh, D. Comparison of hyperemotionality caused by lesions in the septal and ventromedial hypothalamic areas in the rat. *Psychonomic Science,* 1969, **16,** 3-4.

Singh, D. Preference for mode of obtaining reinforcement in rats with lesions in septal or ventromedial hypothalamic area. *Journal of Comparative and Physiological Psychology,* 1972, **80,** 259-268.

Singh, D., & Meyer, D.R. Eating and drinking by rats with lesions of the septum and the ventromedial hypothalamus. *Journal of Comparative and Physiological Psychology*, 1968, **65**, 163-166.

Smith, C.J.V., & Britt, D.L. Obesity in the rat induced by hypothalamic implants of goldthioglucose. *Physiology and Behavior*, 1971, **7**, 7-10.

Smith, H.C., & Boyarsky, S. The relationship between physique and simple reaction time. *Character and Personality*, 1943, **12**, 46-53.

Smith, M., & Duffy, M. Some physiological factors that regulate eating behavior. *Journal of Comparative and Physiological Psychology*, 1957, **50**, 601-608.

Smith, M., Salisbury, R., & Weinberg, H. The reaction of hypothalamic hyperphagic rats to stomach preloads. *Journal of Comparative and Physiological Psychology*, 1961, **54**, 660-664.

Stefanik, P.A., Heald, F.P., & Mayer, J. Caloric intake in relation to energy output of obese and non-obese adolescent boys. *American Journal of Clinical Nutrition*, 1959, **57**, 59-62.

Stellar, E. The physiology of motivation. *Psychological Review*, 1954, **61**, 5-22.

Strominger, J.L., Brobeck, J.R. & Cort, R.L. Regulation of food intake in normal rats and in rats with hypothalamic hyperphagia. *Yale Journal of Biology and Medicine*, 1953, **36**, 55-74.

Stunkard, A.J. Obesity and the denial of hunger. *Psychosomatic Medicine*, 1959, **21**, 281-289.

Stunkard, A., & Koch, C. The interpretation of gastric motility: I. Apparent bias in the resports of hunger by obese persons. *Archives of General Psychiatry*, 1964, **11**, 74-82.

Szasz, T.S. *Pain and pleasure: A study of bodily feelings*. New York: Basic Books, 1957.

Teitelbaum, P. Sensory control of hypothalamic hyperphagia. *Journal of Comparative and Physiological Psychology*, 1955, **48**, 156-163.

Teitelbaum, P. Random and food-directed activity in hyperphagic and normal rats. *Journal of Comparative and Physiological Psychology*, 1957, **50**, 486-490.

Teitelbaum, P., & Campbell, B.A. Ingestion patterns in hyperphagic and normal rats. *Journal of Comparative and Physiological Psychology*, 1958, **51**, 135-140.

Thomas, D.W., & Mayer, J. Meal taking and regulation of food intake by normal and hypothalamic hyperphagic rats. *Journal of Comparative and Physiological Psychology*, 1968, **66**, 642-653.

Thorndike, E.L., & Lorge, I. *The teacher's wordbook of thirty thousand words*. New York: Teachers College, Columbia University, 1944.

Turner, S.G. Sechzer, J.A., & Liebelt, R.A. Sensitivity to electric shock after ventromedial hypothalamic lesions. *Experimental Neurology*, 1967, **19**, 236-244.

Wheatley, M.D. The hypothalamus and affective behavior in cats. *Archives of Neurology and Psychiatry*, 1944, **52**, 296-316.

Williams, D.R., & Teitelbaum, P. Some observations on the starvation resulting from lateral hypothalamic lesions. *Journal of Comparative and Physiological Psychology*, 1959, **52**, 458-465.

Witkin, H.A., Lewis, H.B., Hertzman, M., Machover, K., Meissner, P.B., & Wapner, S. *Personality through perception*. New York: Harper, 1954.

Woodworth, R.S., & Schlosberg, H. *Experimental psychology*. New York: Holt, Rinehart & Winston, 1954.

Wooley, O.W. Long-term food regulation in the obese and nonobese. *Psychosomatic Medicine*, 1971, **33**, 436-444.

Wooley, O.W., Wooley, S.C., & Dunham, R.B. Can calories be perceived and do they affect hunger in obese and nonobese humans? *Journal of Comparative and Physiological Psychology*, 1972, **80**, 250-258.

Wooley, S.C. Physiological vs. cognitive factors in short term in the obese and nonobese. *Psychosomatic Medicine*, 1972, **34**, 62-68.

Zajonc, R.B., Social facilitation, *Science*, 1965, **149**, 269-274.

AUTHOR INDEX

Numbers in *italics* refer to the pages on which the complete references are listed.

SUBJECT INDEX

For Product Safety Concerns and Information please contact our EU representative GPSR@taylorandfrancis.com Taylor & Francis Verlag GmbH, Kaufingerstraße 24, 80331 München, Germany

Batch number: 10725802

Printed by Printforce, the Netherlands